CIRIA, C750 )16



# Groundwater control: design and practice, second edition

*M Preene Golder Associates (UK) Limited*

*T O L Roberts WJ Groundwater Limited*

*W Powrie University of Southampton*

ciria

Griffin Court, 15 Long Lane, London, EC1A 9PN

Tel: 020 7549 3300 Fax: 020 7549 3349

Email: enquiries@ciria.org Website: www.ciria.org

# Summary

This publication provides information and guidance on pumping methods used to control groundwater as part of the temporary works for construction projects. Subjects covered include potential impact of groundwater on construction works, groundwater control techniques, safety, management and contractual matters, legal and environmental issues that arise when groundwater is pumped and discharged, site investigation requirements, and design methods for groundwater control schemes.

The guide explains the principles of groundwater control by pumping, and gives practical information for the effective and safe design, installation and operation of such works.

This revised publication was prepared under contract to CIRIA by Golder Associates (UK) Limited with support from WJ Groundwater Limited and the University of Southampton. The authors of the original guide (CIRIA C515) were M Preene, T O L Roberts, W Powrie, M R Dyer.

**Groundwater control: design and practice, second edition**

Preene, M, Roberts, T O L, Powrie, W

*CIRIA*

C750 RP990 © CIRIA 2016 ISBN: 978-0-86017-755-5

**British Library Cataloguing in Publication Data**

A catalogue record is available for this book from the British Library

<table>
<tr><td colspan="2">Keywords<br>Groundwater control, dewatering, case histories, contractual aspects, design and operation, environmental matters, excavations, ground engineering, investigation, pore water pressure, pumping, regulations, temporary works</td></tr>
<tr><td>Reader interest<br>Civil and geotechnical engineers, temporary works designers and planners involved in investigation, design, specification, installation, operation and supervision for projects where groundwater control may be required</td><td>Classification<br>Availability Unrestricted<br>Content Advice/guidance<br>Status Committee-guided<br>User Civil and geotechnical engineers, construction professionals</td></tr>
</table>



*Cover photograph courtesy Oliver Vincent*

# Acknowledgements

This second edition is an output from CIRIA's ground engineering research programme and supersedes the first edition CIRIA C515 published in 2000.

## Authors

### *Dr Martin Preene BEng PhD CEng FICE CGeol FGS CEnv CSci C.WEM FCIWEM*

Martin Preene is a dewatering specialist and groundwater engineer with 30 years' experience in the management, control and exploitation of groundwater. He has worked for contractors and consultants on civil engineering, mining, oil and gas and infrastructure projects worldwide. He provides dewatering consultancy services at Preene Groundwater Consulting Limited.

Dr Preene is a UK Registered Ground Engineering Advisor, Chartered Engineer, Chartered Geologist, Chartered Water and Environmental Manager and Chartered Environmentalist. He has wide professional interests in groundwater and hydrogeology and is the author of more than 50 groundwater publications, including a dewatering textbook and several industry guidance documents on the investigation and control of groundwater.

### *Dr Toby Roberts FREng PhD CEng CGeol FICE FGS*

Toby Roberts is an expert on the design and implementation of groundwater control systems for large excavations, foundations and tunnels with extensive contract experience in the Middle East and Europe, and has acted as advisor for major international projects. Toby is a founder member and current chairman of the international dewatering contractor WJ Groundwater Ltd. He has particular expertise on recharge systems, in-tunnel well drilling techniques, design of pumping tests, groundwater remediation and treatment, and has published widely at international conferences and in journals on dewatering methods and technology. Dr Toby Roberts is a Fellow of the Royal Academy of Engineering, a Chartered Civil Engineer and Chartered Geologist.

### *Professor W Powrie FREng MA MSc PhD FICE CEng*

William Powrie is Professor of geotechnical engineering and Dean of the Faculty of Engineering and the Environment at the University of Southampton. His main technical areas of expertise are in geotechnical aspects of transport infrastructure, and sustainable waste and resource management.

William's work on geotechnical aspects of transport infrastructure encompasses groundwater control, retaining walls, earthworks, railway track and fundamental soil behaviour. He is a former associate editor of the *Canadian Geotechnical Journal*, a former honorary editor of the Institution of Civil Engineers journals *Geotechnical Engineering* and *Waste and Resource Management*, and author of the widely respected and best-selling textbook, *Soil mechanics: concepts and applications*, now in its third edition. He was elected Fellow of the Royal Academy of Engineering in 2009.

## Project advisors and consultees

The research project was guided by an advisory group who reviewed CIRIA C515 (the first edition) and who then recommended the scope for this update. The advisory group comprised:

| | |
|---|---|
| C P Chiverrell | CIRIA |
| W Powrie | University of Southampton |
| M Preene | Preene Groundwater Consulting Limited |
| T O L Roberts | WJ Groundwater Limited |
| David Seccombe | Environment Agency |

In addition to the advisory group, the technical contribution of the following specialists is gratefully acknowledged:

David Hartwell (Chapter 2)

David Seccombe (Chapter 4)

Jim White (Chapter 6)

## CIRIA Project manager

Chris Chiverrell

## Project funders

The research project was funded by WJ Groundwater Limited and Golder Associates (UK) Ltd.

## Other acknowledgements

The authors of the first edition of the guide were:

| | |
|---|---|
| Dr M R Dyer | Trinity College, Dublin (formerly Mark Dyer Associates) |
| Professor W Powrie | University of Southampton |
| Dr M Preene | Preene Groundwater Consulting Limited (formerly WJ Groundwater Limited) |
| Dr T O L Roberts | WJ Groundwater Limited |

# Contents

## *Boxes*

### *Case studies*



### *Figures*

### *Tables*

# Glossary

**Analytical model** A simple mathematical model describing an aquifer and its boundary conditions.

**Anisotropy** The condition in which one or more of the properties of an aquifer varies according to the direction of measurement.

**Aquiclude** Soil or rock forming a stratum, group of strata or part of a stratum of very low permeability, which acts as a barrier to groundwater flow.

**Aquifer** Soil or rock forming a stratum, group of strata or part of a stratum that is water-bearing (ie saturated and relatively permeable).

**Aquitard** Soil or rock forming a stratum, group of strata or part of a stratum of intermediate to low permeability, which only yields very small groundwater flows.

**Artificial recharge** Replenishment of groundwater artificially (via wells, pits or trenches) to reduce drawdowns external to a groundwater control system or as a means to dispose of the discharge. (Note that in the UK permission may be required from the regulator to allow artificial recharge, see Chapter 4).

**Barrier boundary** An aquifer boundary that is not a source of water.

**Base heave** Lifting of the floor of an excavation caused by unrelieved pore water pressures.

**Biofouling** Clogging of wells, pumps or pipework as a result of bacterial growth.

**Borehole** A hole drilled into the ground for any purpose, including site investigation boreholes. In groundwater terminology a borehole is often taken to mean a relatively small diameter well, which may or may not have a pump installed.

**Capillary saturated zone** The zone that may exist above the phreatic surface in a fine-grained unconfined aquifer when the soil remains saturated at negative (ie less than atmospheric) pore water pressures.

**Cavitation** The formation of vapour bubbles in water when the static pressure falls below the vapour pressure of water (which can occur inside certain types of pumps and ejectors). When the bubbles move to areas of higher pressure they may implode, causing shock waves that can damage the internal components of pumps and ejectors.

**Cofferdam** A temporary retaining wall structure, which may also exclude lateral flows of groundwater and surface water from an excavation.

**Confined aquifer** An aquifer overlain by a confining stratum of significantly lower permeability than the aquifer and where the piezometric level is above the base of the confining stratum (as a result the aquifer is saturated throughout). Also known as sub-artesian aquifer.

**Consolidation settlements** Ground settlements resulting from a reduction in groundwater levels or piezometric level and the resulting increase in vertical effective stress.

**Constant head test** A form of *in situ* permeability test carried out in boreholes or piezometers where water is added to or removed from the borehole.

The water is maintained at a constant level and the flow rate into or out of the borehole is monitored.

**Construction dewatering** Groundwater control.

**Deepwell** A groundwater extraction well of sufficient dimensions to accept a submersible pump.

**Deepwell pump** Slimline electric submersible pump designed to be used in deepwells. Also known as borehole pump.

**Dipmeter** A portable device for measuring the depth to water in a borehole, well, piezometer or standpipe.

**Discharge** The flow rate pumped by a groundwater control system.

**Discharge permission** Permission from the regulatory authorities to allow water to be discharged from site.

**Drawdown** The amount of lowering of the water table in an unconfined aquifer or of the piezometric level in a confined aquifer caused by a groundwater control system.

**Ejector** A water jet pump which creates a vacuum by circulating clean water at high pressure through a nozzle and venturi arrangement located in a well. Also known as an eductor.

**Electro-osmosis** A groundwater control method used in very low permeability soils where an electric potential difference is applied to the ground to induce groundwater flow.

**Falling head test** A form of *in situ* permeability test carried out in boreholes or piezometers where water is added to raise the water level in the borehole, and the rate at which the water level falls is monitored.

**Filter pack** Sand or gravel placed around a well screen to stabilise the aquifer and to act as a filter and to control movement of fine particles from the surrounding soil.

**Formation level** The final dig level of an excavation.

**Formation stabiliser** A coarse permeable filter gravel placed around a well screen in conditions when there is no requirement to act as a filter. The gravel acts as a permeable backfill to prevent aquifer material from collapsing against and distorting the well screen.

**French drain** A gently sloping drain consisting of a perforated pipe with gravel surround.

**Groundwater** Water contained within and flowing through the pores and fabric of soil and fissures in rock. In hydrogeological terminology, strictly refers to the water within the saturated aquifer or perched aquifers.

**Groundwater control system** A system used to manipulate groundwater levels and flows usually to facilitate construction works. Schemes may involve use of wells, drains, sumps or cut-offs individually or in combination.

**Hazen's formula** An empirical method that can be applied to particle size distributions to estimate approximate permeability values for samples of uniform sands.

**Hydraulic gradient** The change in total hydraulic head between two points, divided by the length of flow path between the points.

**Hydrogeology** The study of the interrelationships of the geology of soils and rock with groundwater. Also known as groundwater hydrology or, especially in the USA, as geohydrology.

**Leaky aquifer** An aquifer confined by a low permeability aquitard. When the aquifer is pumped, groundwater may flow from the aquitard and recharge the aquifer. Also known as a semi-confined aquifer.

**Loss of fines** The movement of clay, silt or sand-size particles out of a soil toward a sump or well where filters are absent or inadequate. (Also describes the washing of finer particles out of a granular soil sample recovered from a borehole during cable percussion drilling.)

**Numerical model** A groundwater flow model where the aquifer and boundary conditions are described by equations and are solved numerically by computer, often by iteration.

**Observation well** A well or borehole used for monitoring groundwater levels or piezometric head.

**Overbleed** Residual groundwater seepage trapped above a lower permeability stratum. See also *Perched water.*

**Overflowing artesian well** A well penetrating a confined aquifer that will overflow naturally without the need for pumping (for this to occur the piezometric level in the aquifer must be above ground level at the well location).

**Packer test** A form of *in situ* permeability test typically carried out in an unlined borehole in rock where a section of borehole is sealed off by inflatable packers and water is pumped into or out of the test section.

**Particle size distribution** The relative percentages by dry weight of particles of different sizes, determined in the laboratory, for a soil sample. Also known as PSD, soil grading, sieve analysis.

**Perched water** Water in an isolated saturated zone above the water table. It is the result of the presence of a layer of low or very low permeability above which water can pond. See also *Overbleed.*

**Permeability** A measure of the ease with which water can flow through the pores of soil or rock. Also known as coefficient of permeability, hydraulic conductivity.

**Phreatic surface** The level at which the pore water pressure is zero (ie atmospheric). Also known as phreatic level. See also *Water table.*

**Physical cut-off** An artificial barrier to groundwater flow, such as a sheet-pile wall or a grout curtain, which excludes or restricts groundwater flows.

**Piezometer** An instrument installed into a soil or rock stratum for monitoring the groundwater level, piezometric level or pore water pressure at a specific point.

**Piezometric level** The level representing the total hydraulic head of groundwater in a confined aquifer. Also known as piezometric surface.

**Plane flow** A two-dimensional flow regime in which flow occurs in a series of parallel planes (eg perpendicular to a pumped slot).

**Pore water pressure** The pressure of groundwater in a soil, measured relative to atmospheric pressure.

**Pumping test** A form of *in situ* permeability test involving pumping from a well (or borehole) and recording the flow rate from the pumped well and groundwater level changes in observation wells and pumped well.

**Radial flow** A two-dimensional flow regime in which flow occurs in planes which converge on an axis of radial symmetry (eg a pumped well).

**Radius of influence** The distance outward from a well or groundwater control system to which the drawdown resulting from pumping extends. Also known as distance of influence.

**Recharge boundary** A boundary that can act as a supply of water to the aquifer. Examples include a more permeable saturated stratum or a watercourse or pond that is in hydraulic connection with the aquifer. A groundwater control system can impact or be affected by the proximity of a recharge boundary.

**Recharge well** A well specifically designed so that water can be pumped into an aquifer. See also *Artificial recharge*.

**Relief well** A well in the base of an excavation which is allowed to overflow in order to relieve pore water pressures at depth. Also known as bleed well.

**Rising head test** A form of *in situ* permeability test carried out in boreholes or piezometers where water is removed to lower the water level in the borehole, and the rate at which the water level rises is monitored.

**Rock** The term used in civil engineering to describe geological deposits formed from mineral grains or crystals cemented together – this is distinct from uncemented soil. Typically in rock the flow of groundwater will be predominantly through fissures or fractures, although intergranular flow can occur in some rock types, and in weathered rock.

**Saturated zone** The part of an unconfined aquifer below the water table where the soil pores are completely filled with water at positive pore water pressures.

**Seasonal variation** Natural variation in groundwater levels during the course of a year.

**Soil** The term used in civil engineering to describe uncemented deposits of mineral (and occasionally organic) particles such as gravel, sand, silt and clay – this is distinct from cemented rock. Typically in soil the flow of groundwater will be predominantly intergranular (ie through the pore spaces between the soil grains).

**Standpipe** An instrument, typically consisting of an open perforated tube, installed into the ground for monitoring the groundwater levels.

**Standpipe piezometer** An instrument, typically consisting of a tube and screen with short response zone, installed into the ground for monitoring the groundwater levels at a defined point.

**Storage coefficient** The quantity of water an aquifer releases per unit surface area of the aquifer per unit drawdown. Also known as storativity.

**Submersible pump** Electric pump commonly used for sump pumping. Slimline pumps are available for use in deepwells. See also *Deepwell pump*.

**Suction lift** The vertical height from the intake of a suction pump to the surface of the water being pumped from a well or sump. Typically this depth is limited to 7 m or less.

**Sump** A pit usually located within an excavation where surface and groundwater are allowed to collect prior to being pumped away.

**Sump pump** A pump, capable of handling solids-laden water, used to pump from sumps.

**Surface water** Water from precipitation, leakage or from lakes, rivers etc, which has not soaked into the ground.

**Tidal variation** Cyclical changes in groundwater level or piezometric level from the influence of tides.

**Total hydraulic head** A measure of the potential energy of water due to its height above a given level. The total head controls the height at which water will stabilise in a piezometer. The total head at a given point is the sum of the elevation head (ie the height of the point above an arbitrary datum) and the pressure head (ie the height of water that would be recorded in a standpipe piezometer with a response zone at the given point). Also known as total head or total hydraulic potential.

**Transmissivity** A measure of the ease with which water can flow through the saturated thickness of an aquifer. Transmissivity is equal to the product of permeability and saturated aquifer thickness.

**Unconfined aquifer** An aquifer, not overlain by a relatively impermeable confining layer, where a water table exists and is exposed to the atmosphere. Also known as water table aquifer.

**Unsaturated zone** The portion of an unconfined aquifer above the water table and above the capillary saturated zone where soil pores may contain both water and air.

**Vadose zone** Unsaturated zone.

**Vibrating wire transducer (VWT)** A type of electronic pressure transducer commonly used together with a datalogger to measure groundwater levels in a standpipe piezometer or observation well. VWTs can also be installed directly in the ground to monitor pore water pressures at a specific point.

**V-notch weir** A thin plate weir typically mounted in a tank. Calibration charts allow the flow rate to be estimated from the height of water flowing over the weir.

**Water Framework Directive (WFD)** An EU Directive that commits the UK to achieve good qualitative and quantitative status of all water bodies, including surface water and groundwater. A groundwater control system must not lead to deterioration (temporary or permanent) of the current water body status.

**Water table** The level in an unconfined aquifer at which the pore water pressure is zero (ie atmospheric). See also *Phreatic surface*.

**Watercourse** Any natural or artificial channel above or below ground through which water flows, such as a river, brook, beck, ditch, mill stream or culvert.

**Well** A hole sunk into the ground for the purposes of abstracting water. Wells for groundwater control purposes are generally categorised by their method of pumping as deepwells, ejector wells, or wellpoints. In water supply terminology, a well is often taken to mean a large diameter shaft, as may be dug by hand in developing countries. A smaller diameter well, constructed by a drilling rig, is termed a borehole.

**Well casing** The unperforated section of the well liner, installed at depths where any groundwater present is to be excluded from the well. Also known as plain casing.

**Well development** The process of maximising well yields by removing drilling residue and fine particles from the well, and from the aquifer immediately around the well prior to installation of the pumping equipment.

**Well liner** A generic term for well casing and well screen.

**Well loss** The head loss at a well associated with the flow of groundwater from the aquifer into the well.

**Wellpoint** Small diameter shallow well normally installed at close centres by jetting techniques.

| | |
|---|---|
| **Wellpoint pump** | A pump capable of applying a vacuum to the headermain of a wellpoint system and also of pumping the discharge water away. |
| **Well screen** | The perforated or slotted portion of the well liner in a well, wellpoint or sump. |
| **Yield** | The flow rate from an individual well. Also known as well yield. |

# Abbreviations and acronyms

| | |
|---|---|
| **AGS** | Association of Geotechnical and Geoenvironmental Specialists |
| **AMF** | Automatic mains failure |
| **CDM** | Construction (Design and Management) Regulations |
| **bgl** | Below ground level |
| **BDA** | British Drilling Association |
| **DoE** | Department of the Environment |
| **EA** | Environment Agency |
| **EU** | European Union |
| **GBR** | General Binding Rules |
| **gwl** | Groundwater level |
| **HDPE** | High density polyethylene |
| **HIA** | Hydrogeological impact appraisal |
| **HSE** | Health and Safety Executive |
| **ICE** | Institution of Civil Engineers |
| **IChemE** | Institution of Chemical Engineers |
| **i.d.** | Internal diameter |
| **JCT** | Joint Contracts Tribunal |
| **LNAPL** | Light non-aqueous phase liquid |
| **NEC** | New Engineering Contract |
| **NIEA** | Northern Ireland Environment Agency |
| **NRW** | Natural Resources Wales |
| **o.d.** | Outside diameter |
| **PC** | Personal computer |
| **PCA** | Permitted Controlled Activity |
| **PSD** | Particle size distribution |
| **PVC** | Polyvinyl chloride |
| **SEPA** | Scottish Environment Protection Agency |
| **SPT** | Standard penetration test |
| **SPZ** | Source protection zone |
| **TBM** | Tunnel boring machine |
| **VWT** | Vibrating wire transducer |
| **WFD** | Water Framework Directive |

# Notation

| | |
|---|---|
| $A$ | Area |
| $a$ | Length of groundwater control system |
| $B$ | Partial penetration factor for wells |
| $b$ | Width of equivalent slot; |
| | Width of groundwater control system |
| | Half width of cofferdam |
| $C$ | Calibration factor |
| $c_{hv}$ | Coefficient of consolidation for vertical compression of soil under horizontal drainage |
| $c_v$ | Coefficient of consolidation of soil |
| $D$ | Thickness of confined aquifer |
| | Thickness of compressible layer |
| $D_{10}$ | Sieve aperture through which 10 per cent of a soil sample will pass |
| $D_{15}$ | Sieve aperture through which 15 per cent of a soil sample will pass |
| $D_{40}$ | Sieve aperture through which 40 per cent of a soil sample will pass |
| $D_{50}$ | Sieve aperture through which 50 per cent of a soil sample will pass |
| $D_{60}$ | Sieve aperture through which 60 per cent of a soil sample will pass |
| $D_{85}$ | Sieve aperture through which 85 per cent of a soil sample will pass |
| $d$ | Depth to water table |
| | Depth of excavation in cofferdam |
| | Drainage path length |
| $E$ | Young's modulus of soil |
| $E'o$ | Stiffness of soil in one-dimensional compression |
| $G$ | Shape factor for flow to rectangular equivalent wells in confined aquifers |
| | Shear modulus of soil |
| $H$ | Initial groundwater head |
| | Excess head in rising and falling head tests |
| | Applied head in packer test |
| $H_c$ | Excess head in constant head test |
| $H_o$ | Initial head in rising and falling head tests |
| $h$ | Total hydraulic head |
| | Groundwater head |
| | Height of water over weir |
| $h_n$ | Seepage head into a cofferdam |
| $h_w$ | Groundwater head in a pumped well or slot |
| $(H-h)$ | Drawdown |
| $(H-h_w)$ | Drawdown in a pumped well or slot |
| $i$ | Hydraulic gradient |
| $i_{crit}$ | Critical seepage gradient for excavations |
| $i_{max}$ | Maximum hydraulic gradient at entry to a well |
| $k$ | Coefficient of permeability |

| | |
|---|---|
| $k_h$ | Coefficient of permeability in the horizontal direction |
| $k_v$ | Coefficient of permeability in the vertical direction |
| $L$ | Length of test section in packer test |
| $L_o$ | Distance of influence for plane flow |
| $l$ | Cut-off wall penetration below excavation level |
| $l_w$ | Wetted length of well screen |
| $m$ | Seepage factor |
| $m_v$ | Coefficient of volume compressibility of soil |
| $n$ | Number of wells |
| $P$ | Depth of penetration into aquifer of partially penetrating well or slot |
| $Q$ | Flow rate |
| | Flow rate from a groundwater control system |
| $Q_{fp}$ | Flow rate from a fully penetrating well or slot |
| $Q_{pp}$ | Flow rate from a partially penetrating well or slot |
| $q$ | Flow rate from a well |
| $R_o$ | Radius of influence for radial flow |
| $r$ | Radial distance from well |
| | Radius of borehole |
| $r_e$ | Equivalent radius of groundwater control system |
| $r_w$ | Radius of well |
| $S$ | Groundwater storage coefficient |
| $s$ | Drawdown |
| $s_o$ | Drawdown imposed in the soil immediately adjacent to a line of wells |
| $T$ | Transmissivity |
| | Time factor |
| $T_r$ | Radial time factor |
| $t$ | Elapsed time |
| $U$ | Uniformity coefficient |
| $u$ | Pore water pressure |
| | Argument of Theis well function |
| $W(u)$ | Theis well function |
| $x$ | Linear distance |
| | Length of pumped slot |
| $\zeta$ | Depth |
| $\alpha$ | V-notch angle of weir |
| $\gamma_s$ | Unit weight of soil |
| $\gamma_w$ | Unit weight of water |
| $\lambda$ | Partial penetration factor for confined slots |
| $\nu'$ | Poisson's ratio |
| $\rho$ | Vertical settlement |
| $\sigma$ | Total stress |
| $\sigma'$ | Effective stress |
| $\sigma'_v$ | Vertical effective stress |
| $\tau$ | Shear stress |
| $\phi'$ | Soil angle of shearing resistance |

# 1 Groundwater in construction

## 1.1 INTRODUCTION AND USER GUIDE

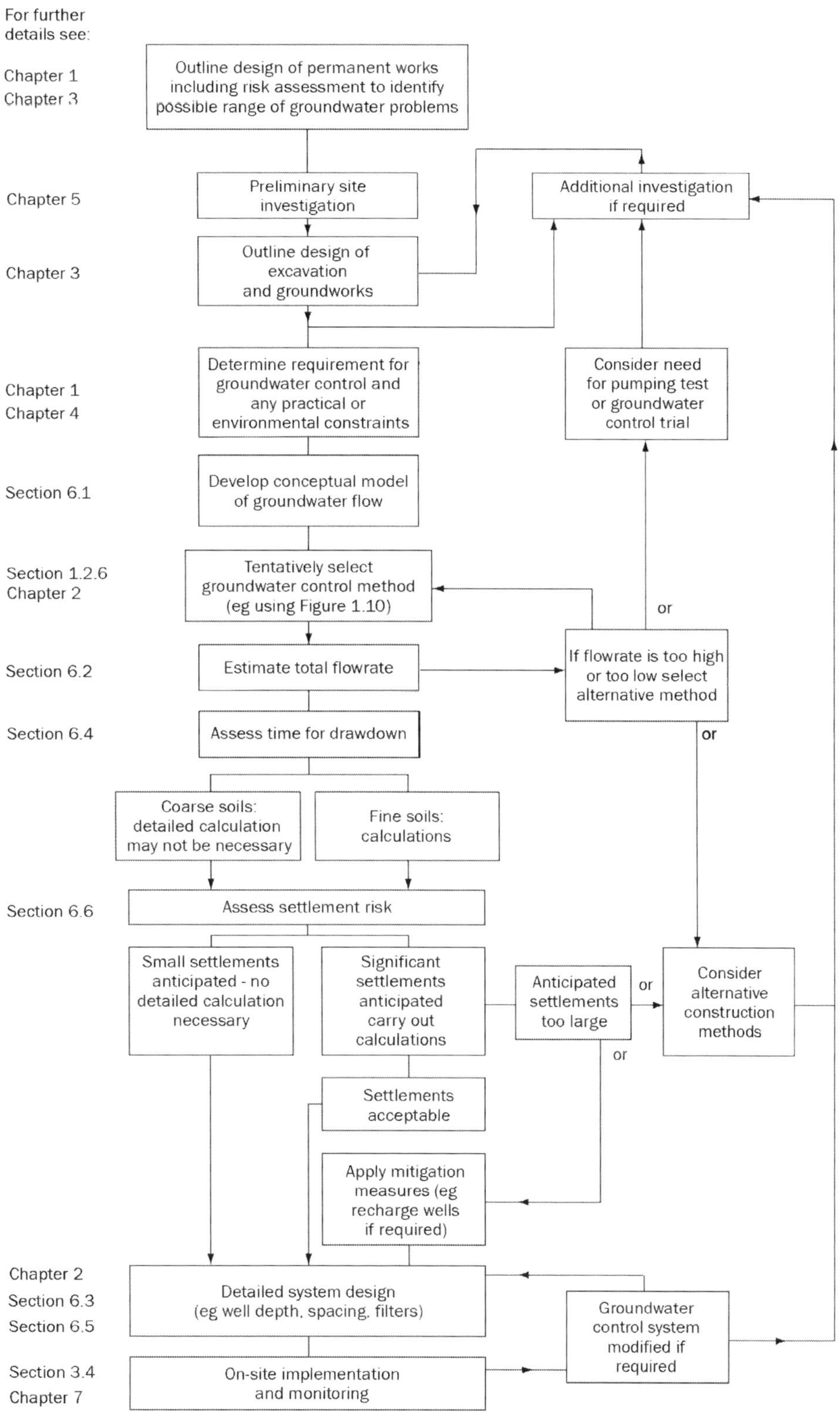


*Figure 1.1* ***Principal stages in the analysis and design of groundwater control systems***

Whenever an excavation is made below the natural water table, there is a risk that it will become unstable or flood unless steps are taken to control the groundwater in the surrounding soil (see Figure 1.2). Groundwater may be controlled by installing a physical barrier to exclude groundwater from the excavation, or by pumping groundwater from specially installed wells in order to lower artificially the water table in the vicinity of the excavation, or by a combination of the two techniques. Often, the use of a pumped well system, either alone or in combination with a physical barrier, will be the most economical and convenient approach. The appropriate type of pumped well system to use depends primarily on the nature of the ground and the depth of the excavation.

This guide explains the design and operation of groundwater control systems involving pumping from wells. It is divided into the following chapters:

**Chapter 1:** technical principles of groundwater flow and control

**Chapter 2:** commonly used methods of groundwater control

**Chapter 3:** management of pumped well groundwater control systems

**Chapter 4:** environmental considerations

**Chapter 5:** site investigation

**Chapter 6:** methods of analysis and design

**Chapter 7:** case studies.

The number of excavations where no consideration need be given to the potential effects of groundwater is very small. Thus the design, installation and operation of a groundwater control system – and obtaining the necessary site investigation data – should be viewed as an integral part of the overall works.

### 1.1.1 Users

The guidance given in this report is intended for use by those concerned with the design, specification, installation, operation, monitoring or management of pumped well groundwater control systems. As such it is intended to be accessible at a number of levels as:

- background information for project managers, resident engineers, site agents and others who encounter groundwater control systems during the course of their work and need to be able to discuss particular aspects with specialist groundwater contractors or consultants
- an introduction to the subject for geotechnical engineers with little or no previous experience of groundwater control
- a reference or source book for more experienced geotechnical engineers.

Technical details and case histories are presented in boxes, separately from the main text. The report is divided into sections and sub-sections. A feature to help the reader is the extensive cross-referencing between sections (highlighted in the left hand margins). Figure 1.1 shows a flow diagram of the principal stages in analysis and design of groundwater control systems, and the corresponding sections of this guide.

### 1.1.2 Limitations

The guide is a comprehensive, up-to-date guide to the design and operation of pumped well groundwater control systems, but it is not intended to be a 'do-it-yourself' manual on dewatering for the novice. Success in ground engineering usually depends on the application of engineering judgement, which in turn requires not only a thorough understanding of the principles involved, but also a measure of experience. This guide is not a substitute for professional advice: if in doubt, consult an expert.

The guide does not cover exclusion methods of groundwater control, except to list them and indicate where further information may be found.

*a) Slumping of side slopes caused by seepage into an excavation in fine sands*

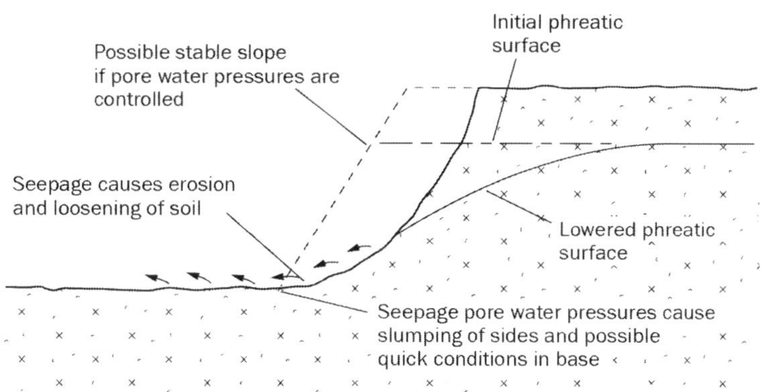


*b) Instability of side slopes*

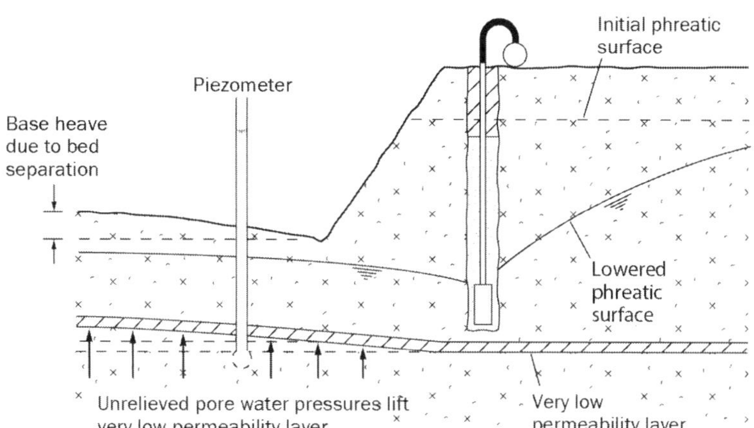


*c) Instability of base due to unrelieved pore water pressures*

*Figure 1.2 Groundwater-induced instability of excavation (from Preene and Powrie, 1994)*

# 1.2 OBJECTIVES AND OVERVIEW OF GROUNDWATER CONTROL

## 1.2.1 Groundwater in the environment

The total volume of water on the earth is large, but finite. Most of it is in constant motion, in what is known as the hydrological cycle (Figure 1.3). Some of the water, which falls on the land as precipitation (rain, hail, sleet or snow) runs off into surface streams, rivers and ponds. Some evaporates directly, and the remainder infiltrates into the ground. A proportion of the water that infiltrates into the ground is taken up by plants through their roots, and the rest moves generally downward through the near-surface zone until it reaches the groundwater level or water table. The study of groundwater is encompassed by the field of hydrogeology. Further guidance can be found in Freeze and Cherry (1979), Price (1996) Fetter (2014), Brassington (2006) and Younger (2007).

The guidance given in this report is primarily aimed at construction projects where excavations are to be made in soil – uncemented deposits of mineral (and occasionally organic) particles such as gravel, sand, silt and clay. The soil particles are in contact with each other, but with voids in between them. These voids are known as soil pores, and flow of groundwater in soil is predominantly through the soil pores. Many of the techniques described in this report are also relevant to excavations in rock – deposits formed from mineral grains or crystals cemented together. Typically in rock the flow of groundwater will be predominantly through fissures or fractures, although intergranular flow can occur in some rock types, and in weathered rock.

Water contained in the soil pores (and within fissures and fractures in rock) is known as groundwater. Below the water table, the soil pores are full of water, and the soil is saturated. Above the water table, the soil pores will generally contain both air and water.

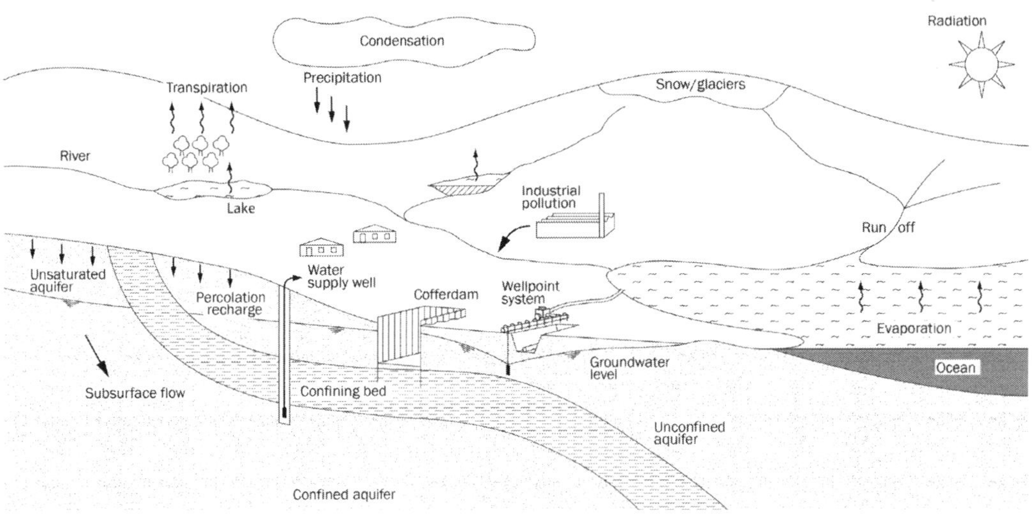


*Figure 1.3 The hydrological cycle*

The balance between the air and water in the zone of soil or rock above the water table is influenced by the pore size or fracture opening. In coarse-grained soils, the voids may contain significant quantities of air, and the soil or rock above the water table will often be unsaturated. Fine-grained materials can retain water in the voids by capillary action, remaining saturated for some height above the water table. The zone of unsaturated soil or rock near the surface is known as the vadose zone.

The pressure of the water in the soil voids at any point is termed the pore water pressure. The pore water pressure is measured relative to atmospheric pressure (ie a pore water pressure of 100 kPa means 100 kPa above atmospheric pressure). The pore water pressure is important because it affects not only the direction and speed of groundwater flow, but also the stability of the soil around or below an excavation (see Sections 1.2.4 and 1.2.5).

In fissured rock the same principles apply, but most of the groundwater that can move freely is contained in the fissures rather than in pores in the intact lumps of rock.

Excavations below the groundwater level are vulnerable to instability, erosion and flooding from the effects of groundwater (Figure 1.2), surface water and, in extreme cases, precipitation. This report is concerned with the protection of excavations below the water table from the effects of groundwater alone, and of groundwater and surface water acting in combination (eg where a stream or river acts as a source of recharge to the groundwater). This guide does not deal with the preventive measures used to protect excavations from the direct effects of surface water or precipitation.

## 1.2.2 Aquifers, aquicludes and aquitards

**See also**
1.2.4 Permeability
Box 6.3 Aquifers

Water can flow much more readily through the pores in coarse-grained soils (eg gravels and coarse sands) and fissures in rocks than through the pores in fine-grained soils (eg silts and clays). The ease with which water can flow through the pores of a soil or rock is expressed in terms of the permeability or hydraulic conductivity (Section 1.2.4).

Soils and rocks of high permeability whose voids are full of water are termed aquifers, while soils and rocks of such low permeability that they act as a seal are termed aquicludes. Strata of intermediate permeability, relative to aquifers and aquicludes, and which allow water to flow through them but only slowly, are termed aquitards. Usually, pumped well systems are used to control groundwater during temporary works in soils that are either aquifers or aquitards.

If the upper surface of an aquifer is exposed to the atmosphere, the aquifer is known as an unconfined or water table aquifer. If, on the other hand, the aquifer is fully saturated and overlain by a comparatively impermeable stratum or aquitard, the aquifer is described as confined. These terms are illustrated in Box 1.1 (see also Box 6.3).

## 1.2.3 Natural pore water pressures in the ground

The natural pore water pressures in the ground at a site depend on the ground conditions and the natural groundwater flow regime. The water table (or phreatic surface) may be defined as the level at which the pore water pressure (measured relative to atmospheric pressure) is zero. If the groundwater is at rest (or flowing horizontally through a uniform aquifer), the pore water pressures will be hydrostatic (Box 1.2).

*Box 1.1 Non-hydrostatic groundwater conditions*

An aquifer overlain by a clay soil in a river valley is shown below. The aquifer extends beyond the edges of the clay, up into the surrounding hills. In the valley where the aquifer is overlain by the clay the aquifer is confined; in the hills where its surface is exposed to the atmosphere the aquifer is unconfined. The pore water pressures in the aquifer where it is confined in the valley can be high, because the pore water can flow relatively easily through the aquifer from the high hills while the clay acts as a seal. A standpipe driven through the clay may indicate a water level or piezometric level in the aquifer which is above the ground surface in the valley. If the standpipe is not tall enough it will overflow, bringing water from the aquifer to the surface. At the ground surface, the pore water pressure is zero. At the base of the clay layer, the pore water pressure is equal to the unit weight of water $\gamma_w$ multiplied by the height to which the water rises in the standpipe (assuming that it is tall enough to prevent overflowing). The pore water pressures in the aquiclude are greater than they would be if the groundwater conditions were hydrostatic below a water table at the ground surface. Groundwater flows upward through the clay, but probably not more quickly than it can evaporate from the ground surface.

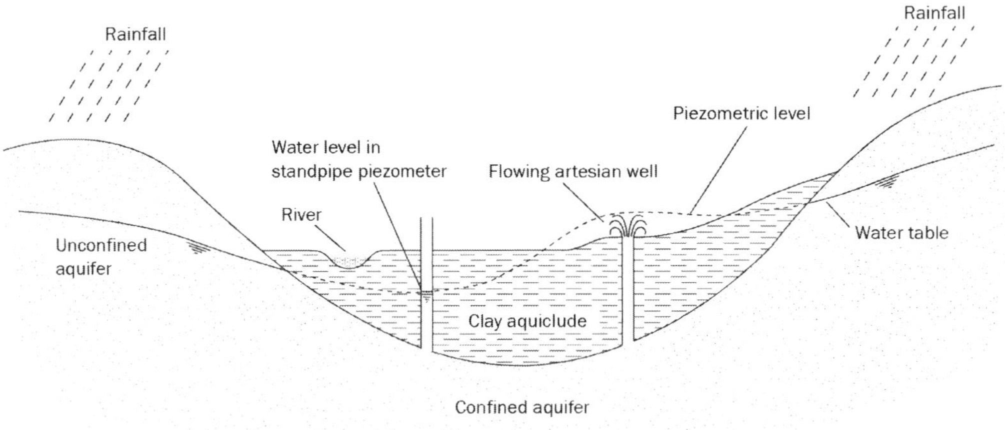


***Cross-section through confined and unconfined aquifers with flowing artesian groundwater conditions***

*Box 1.2 Hydrostatic groundwater conditions*

If the groundwater is at rest (or flowing horizontally through a single, uniform stratum), the pore water pressures will be hydrostatic below the water table; ie at a depth *z*, the pore water pressure (in kPa) will be equal to the unit weight of water $\gamma_w$ (in kN/m$^3$) multiplied by the depth below the water table ($z-d$) (in m). In the vicinity of an excavation where pumping is being carried out or where there is a significant vertical flow of groundwater, the increase in pore water pressure with depth will not in general be hydrostatic.

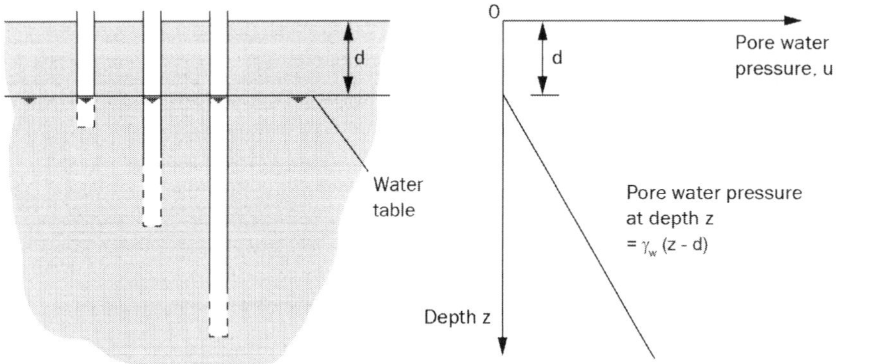


***Hydrostatic pore water pressure distribution***

Non-hydrostatic conditions are usually associated with significant vertical groundwater flow. One example of this is when the pore water pressure in a confined aquifer is high enough to cause water to flow very slowly upward through the overlying aquiclude (Box 1.1). If a well is drilled through the aquiclude to the underlying aquifer, the well will overflow. Such a well is known as a flowing artesian well, and the conditions that cause it are termed artesian or flowing artesian.

In an unconfined aquifer, the pore water pressures above the water table can be negative, rather than positive. There is, however, a limit to the negative pore water pressure a soil can sustain without drawing in air (at atmospheric pressure) through any surface which is exposed to the atmosphere. This limiting negative pore water pressure is known as the air entry value, and increases as the soil pore size decreases. The consequence is that coarse soils above the water table (at which the pore water pressure is zero) will tend to be unsaturated, with very little water retained in the pores by capillary action. Fine-grained soils (ie silts and clays) may remain saturated for several metres above the water table, with pore water pressures continuing to decrease until the air entry value is reached (Figure 1.4).

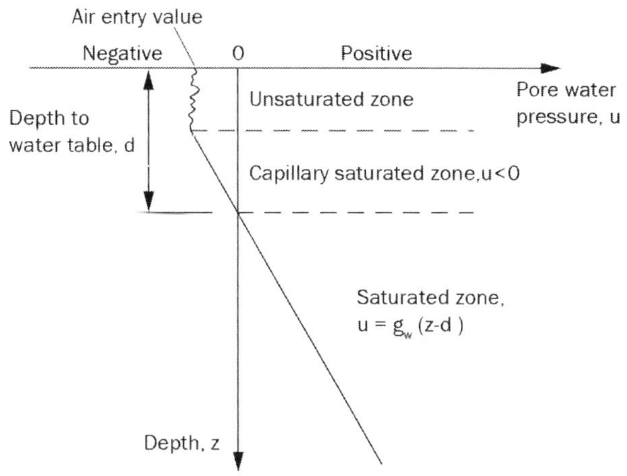


*Figure 1.4* ***Pore water pressures in a fine-grained soil above the water table (groundwater at rest) (after Bolton, 1991)***

## 1.2.4 Groundwater flow and permeability

If the pore water is at rest, the distribution of pore water pressure must be hydrostatic (Box 1.2). Conversely, any localised change in pore water pressure from the hydrostatic value will cause water to flow through the voids between the soil particles. Groundwater flow is driven by a difference in the total hydraulic head, which may be defined as the height to which water rises in a pipe, inserted with its tip at the point where the head is to be measured (Box 1.3). The total hydraulic head may be measured from any convenient datum, but once the datum level has been chosen for a particular situation, it should not be changed. The total hydraulic head is also known as the total head or the hydraulic potential.

In 1836 Robert Stephenson used pumped wells to lower groundwater levels, to enable the construction of the Kilsby tunnel on the London to Birmingham railway, in Northamptonshire (Preene, 2004). Stephenson observed that on pumping from one well, the water levels in adjacent wells dropped. He also recognised that the head difference between the wells was, for a given rate of pumping, an indication of the ease with which water could flow through the soil. In 1856 Henri Darcy, on the basis of a series of experiments carried out at Dijon in France, proposed what is now known as Darcy's Law, which describes the flow of groundwater through saturated soil (Box 1.3).

**See also**
5.3.5 Particle size analysis

The coefficient of permeability used in Darcy's Law is a measure of the ease with which water can flow through the voids between the soil particles, and depends on the properties of the permeant fluid as well as of the soil matrix. For uniform soils, Darcy's coefficient of permeability depends on a number of factors including the void size, the void ratio, the arrangement of particles and the viscosity of the pore fluid (which for water varies by a factor of about two between temperatures of 20°C and 60°C). These factors are discussed in detail by Loudon (1952). In a uniform soil the void size (which is related to particle size) is generally by far the most significant factor. Some empirical correlations between particle size and coefficient of permeability are given in Section 5.3.5.

In this report the term permeability, $k$, is used to mean the coefficient of permeability with water as the permeating fluid, as defined by Darcy's Law (the coefficient of permeability is sometimes also called the hydraulic conductivity).

Approximate permeability values for various types of soil are shown in Table 1.1. Note that the overall range is large and is reinforced by comparing the difference in permeability between gravels and clays (a factor of perhaps $10^{10}$) with the difference in shear strength between high tensile steel and soft clay (about $10^5$).

*Box 1.3 Darcy's Law*

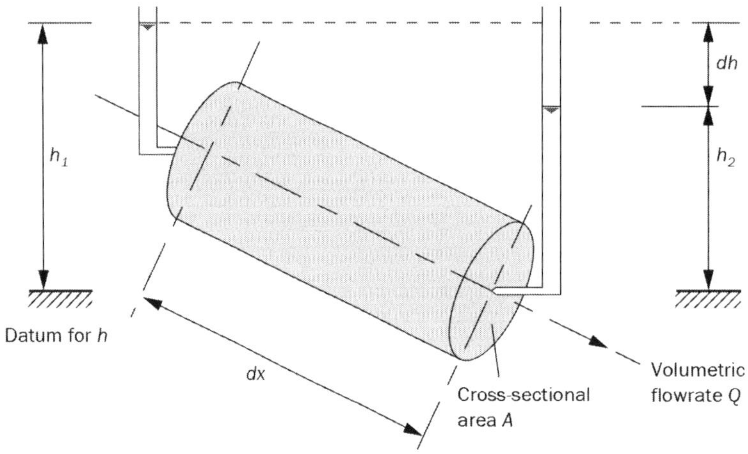


**Darcy's experiment**

Darcy's Law is expressed mathematically as:

$$Q = Aki \tag{1.1}$$

where

$Q$ ($m^3$/s) is the volumetric flow rate of water

$A$ ($m^2$) is the cross-sectional area of flow

$i$ is the rate of decrease of total hydraulic head (potential) $h$ with distance in the direction of the flow ($x$), $-dh/dx$, termed the hydraulic gradient

$k$ (m/s) is a soil parameter known as the coefficient of permeability or the saturated hydraulic conductivity

**Note**

The negative sign in the definition of the hydraulic gradient is mathematically necessary because the flow is always in the direction of decreasing head. If $dh/dx$ is positive, the flow rate will be in the negative $x$ direction. If $dh/dx$ is negative, the flow rate will be in the positive $x$ direction.

The main condition required for Darcy's Law to be valid is that groundwater flow should be laminar, rather than turbulent. In soils which have a particle size larger than a coarse gravel, groundwater velocities may be large enough for turbulent flow. In most other geotechnical applications, flow will be laminar. It is normally assumed that the soil is saturated. The permeability of an unsaturated or a partly saturated soil is an altogether different matter. Surface tension effects offer considerable resistance to flow, so that when a soil becomes unsaturated its permeability will fall by perhaps three orders of magnitude. These effects are discussed by McWhorter (1985).


**See also**

5.3 Permeability testing

6.1.3 Permeability selection


Many analytical methods assume that the ground can be assigned a single value of permeability, which is the same in all directions and does not vary from point to point. In reality, the permeability is likely to be different in the vertical and horizontal directions as a result of deposition-induced anisotropy or layering, and to vary significantly because of heterogeneities such as fissures, sand lenses etc (see Sections 5.3 and 6.1.3). The influence of soil fabric and structure on permeability is discussed by Rowe (1972). The permeability of a confined aquifer $k$ is sometimes multiplied by the saturated thickness of the aquifer $D$ to give a parameter known as the aquifer transmissivity, $T$.

*Table 1.1 Permeabilities of typical soils*

| Indicative soil type | Degree of permeability | Permeability m/s |
|---|---|---|
| Clean gravels | High | $>1 \times 10^{-3}$ |
| Sand and gravel mixtures | Medium | $1 \times 10^{-3}$ to $1 \times 10^{-5}$ |
| Very fine sands, silty sands | Low | $1 \times 10^{-4}$ to $1 \times 10^{-7}$ |
| Silt and interlaminated silt/sand/clays | Very low | $1 \times 10^{-6}$ to $1 \times 10^{-9}$ |
| Intact clays | Practically impermeable | $<1 \times 10^{-9}$ |

## 1.2.5 Groundwater and stability

A saturated soil comprises two phases, the soil particles and the pore water. The strengths of these two phases, in terms of their ability to withstand shear stresses, are very different. The shear strength of water is negligible. The only form of stress that static water can sustain is an isotropic pressure, which is the same in all three principal directions. The soil skeleton, however, can resist shear – mainly because of interparticle friction. The frictional nature of the strength of the soil skeleton means that the higher the normal stress pushing the particles together, the greater the shear stress that can be applied before slip between particles starts to occur.

Because the strengths of the soil skeleton and the pore water are so different, it is necessary to consider the stresses acting on each phase separately. This is achieved by applying the principle of effective stress proposed by Terzaghi in 1936 (Box 1.4).

*Box 1.4 The principle of effective stress*

The effective normal stress $\sigma'$ is the stress carried by the soil skeleton (the soil particles), which controls the volume and strength of the soil. For saturated soils, the effective stress may be calculated from the total normal stress $\sigma$ and the pore water pressure $u$ by Terzaghi's equation:

$$\sigma' = \sigma - u \tag{1.2}$$

As the pore water cannot take shear, all shear stresses must be carried by the soil skeleton.

It is shown in the remainder of this section that pore water pressures have a crucial influence on the stability of the base and sides of an excavation.

### Base stability

A common objective of groundwater control is to maintain the stability of the base and possibly the sides of an excavation. The base of an excavation in a uniform soil will become unstable if the pore water pressure is close to the vertical total stress (due to the weight of the soil), so that the vertical effective stress approaches zero. This condition is known as fluidisation or boiling – quicksand if it occurs over a large area; and piping if it occurs in localised channels.

By considering the forces acting on a block of soil which is on the verge of uplift, it can be shown (see Powrie, 2013) that fluidisation will occur in regions of upward flow in a soil of uniform permeability when the upward hydraulic gradient exceeds a critical value, $i_{crit}$:

$$i_{crit} = (\gamma_s - \gamma_w)/\gamma_w \tag{1.3}$$

where $\gamma_s$ is the unit weight of the soil, and $\gamma_w$ is the unit weight of water (Figure 1.8). For soils with $\gamma_s = 20$ kN/m³ $\approx 2\gamma_w$, then $i_{crit} \approx 1$. The maximum upward hydraulic gradient below the floor of an excavation should not normally exceed $i_{crit}$ .

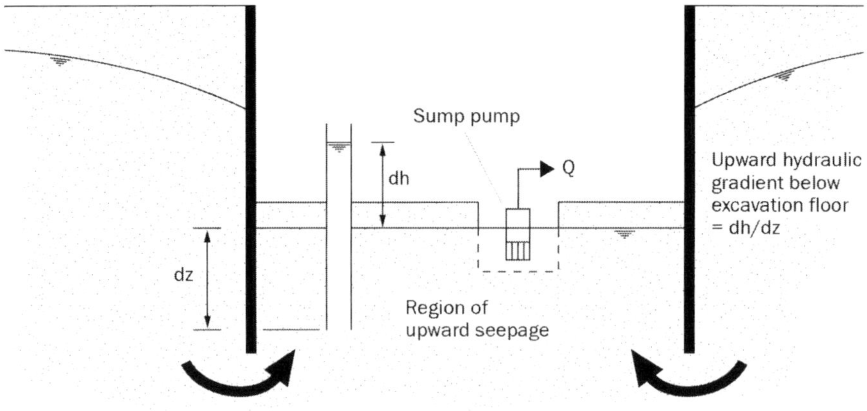


*Figure 1.5 Upward hydraulic gradient for base instability: excavation in a uniform soil*

**See also**
Box 5.1 Base heave

Basal failure or base heave may occur where an excavation is made into a stratum of low permeability soil overlying a confined aquifer (Figure 1.6). Instability is a risk when the upthrust (from the pore water pressure in the confined aquifer) on the base of a plug of the low permeability soil becomes equal to the weight of the soil plug, plus any shear stresses on its sides (see also Hartwell and Nisbet, 1987). A case history illustrating the conditions leading to, and the consequences of, the failure of the base of an excavation is given in Box 1.5 (see also Box 5.1). Instability can be avoided by reducing the pore water pressures in the confined aquifer.

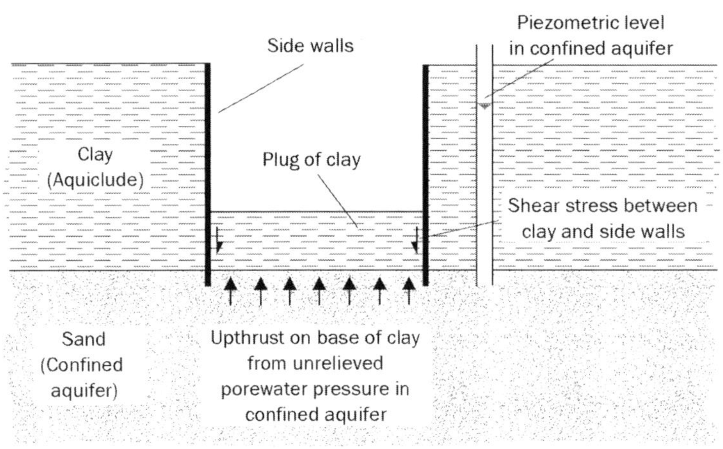


*Figure 1.6 Base failure: excavation in a low permeability soil overlying a confined aquifer*

*Box 1.5* *Case history of base instability in a cofferdam*

A 9 m deep excavation for a pumping station was made in silty clay of estimated permeability $10^{-8}$ m/s underlain by silty sand of estimated permeability $10^{-4}$ m/s. The initial groundwater level in both strata was 1.5 m below ground level. The sides of the excavation were supported by steel sheet-pile retaining walls. To save money, the contractor decided not to install a pumped well system to control the pore water pressures in the silty sand below the base. As the excavation progressed, a point was reached at which the base became unstable and failed, leading to the flooding of the excavation. This resulted in considerable delay and additional cost – concrete props had to be placed underwater to support the retaining walls as the strength of the soil below the floor of the excavation could no longer be relied on, and a pumped well system had to be installed before the excavation could be drained.

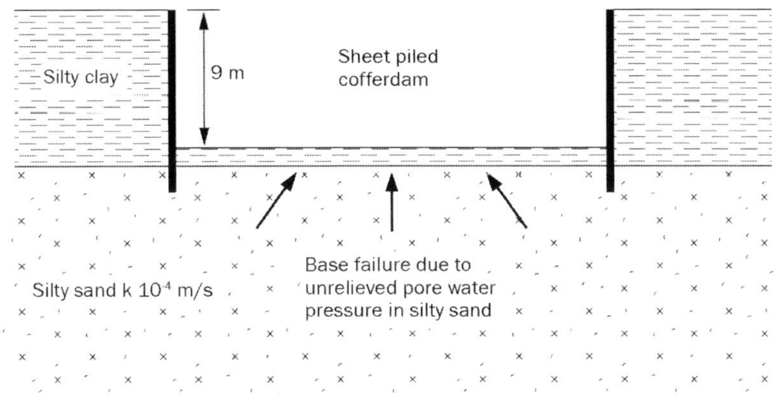


*Cofferdam base instability*

## Side slope stability

**See also**
Figure 1.2 Instability

The shear strength of nearly all soils comes primarily from interparticle friction. The maximum shear stress τ that the soil can resist is directly proportional to the normal effective stress σ′ pushing the soil particles together:

$$\tau = \sigma' \tan\phi' \tag{1.4}$$

where $\phi'$ is the angle of shearing resistance of the soil. Failure will occur when the stress ratio $\tau/\sigma'$ on any plane within the soil mass becomes equal to $\tan\phi'$. Equation 1.4 represents a straight line on a graph of τ against σ′ which defines combinations of shear and normal effective stress at which the soil is at failure. In soil mechanics theory it is known as the Mohr-Coulomb failure criterion.

If a slope is drained, so that the pore water pressure is zero, stable slopes can form at angles equal to the frictional strength of the soil, $\phi'$. If there is seepage out of the slope, it can be shown that the stable angle is reduced to approximately $\phi'/2$ (see Powrie, 2013). In short, lower pore water pressures allow steeper slopes, and seepage flow through slopes reduces the stable angle.

An additional reason for lowering the groundwater level in the vicinity of an excavation is that waterlogged slopes may suffer from erosion if the drawn down water table (also known as the phreatic surface) intersects the cut face of the slope (Figure 1.2a and b and Figure 1.7a). Where a slope cuts through two strata, the lower of which is comparatively impermeable, some overbleed is inevitable (Figure 1.7b). In such cases, the slope should be protected by sandbags, or by the installation of an interceptor drain.

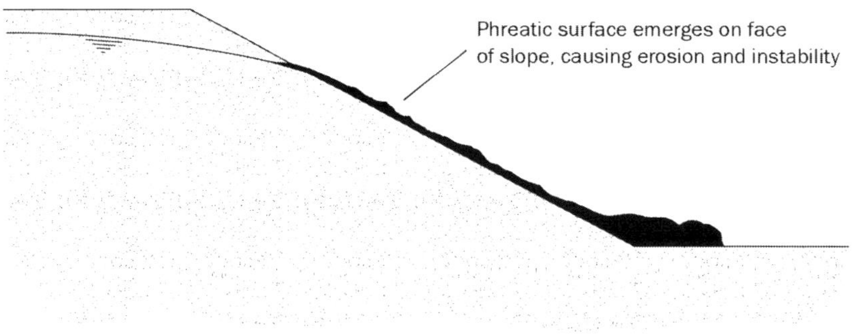


*a) Erosion because of insufficient drawdown*

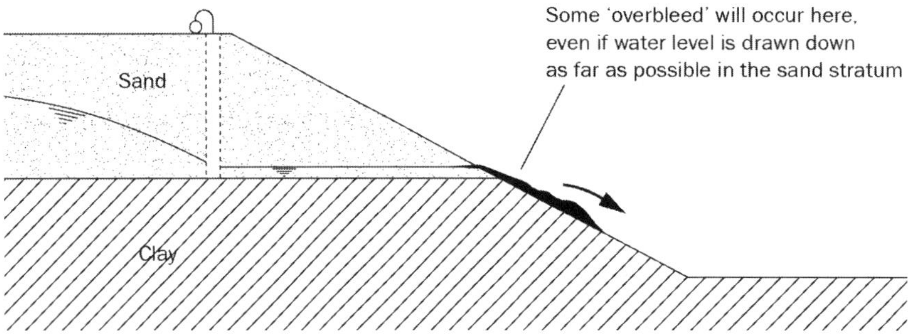


*b) Overbleed on underlying impermeable stratum*

***Figure 1.7   Erosion and overbleed***

## 1.2.6 Objectives of groundwater control

The most obvious (but not necessarily the most important) objective of groundwater control is to prevent an excavation below the natural water table from flooding. Groundwater control can also have an important role in controlling pore water pressures around an excavation to ensure stability of the excavation base and side slopes.

Groundwater control can be achieved by physical exclusion (eg a cut-off wall, ground freezing or grouting), by pumping from sumps or wells (including wellpoints) to intercept the groundwater before it reaches the excavation (resulting in a lowering or drawdown of the water table), or by a combination of the two techniques (Figure 1.8). This guide is concerned with pumped well systems, used either on their own or in combination with a physical cut-off. However, physical cut-offs may be used instead of a pumped well dewatering system, particularly in very coarse-grained and open soils of high permeability.

Where a pumped well dewatering system is installed in an unconfined aquifer, the way in which the required effect (ie a lowering of the water table level) is achieved is subtly different in fine-grained soils and in coarse-grained soils. In a coarse soil the groundwater is able to drain out of the pores in the soil above the water table as the water table is lowered, so that the soil is literally dewatered. Fine-grained soils do not drain freely, so although the level of the water table (defined as the surface of zero pore water pressure) may be lowered, the soil above the new water table will tend to remain saturated. However, the pore water pressure in the soil above the new water table is negative, which increases the effective stress and helps to maintain the stability of the sides or base of an excavation. Strictly speaking, the term dewatering can only be used in connection with unconfined aquifers consisting of coarse-grained soils. For unconfined aquifers consisting of fine-grained soils and confined aquifers, the term pore water pressure control is more appropriate and should therefore be used.

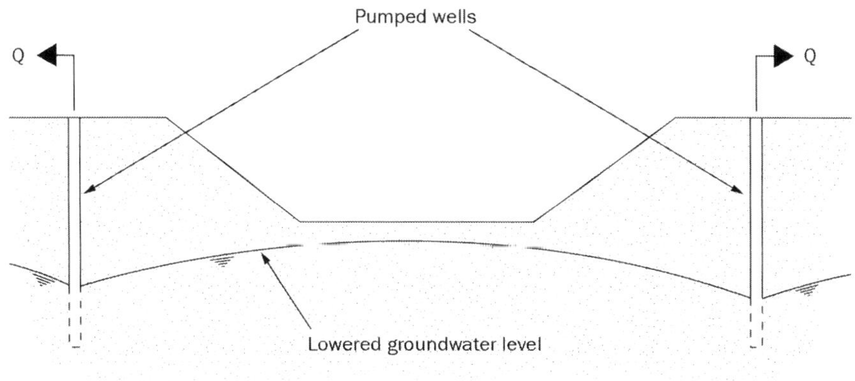


*a) Excavation with battered slopes and external wells*

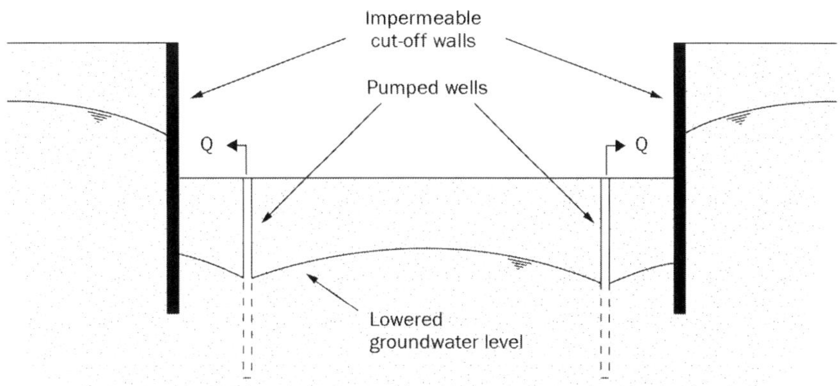


*b) Excavation with a retaining wall and wells to prevent water ingress through the base*

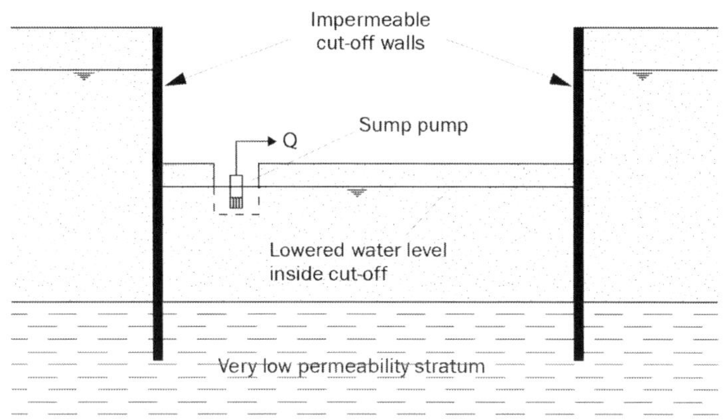


*c) Excavation completely protected by a physical cut-off wall (retaining walls toeing into an impermeable stratum*

*Figure 1.8 Groundwater control using wells and physical cut-offs*

## 1.2.7 Selection of groundwater control method

The principal types of physical cut-off methods are summarised in Table 1.2, and their approximate ranges of application are given in Figure 1.9. Further details of physical cut-off techniques are given by Bell and Mitchell (1986).

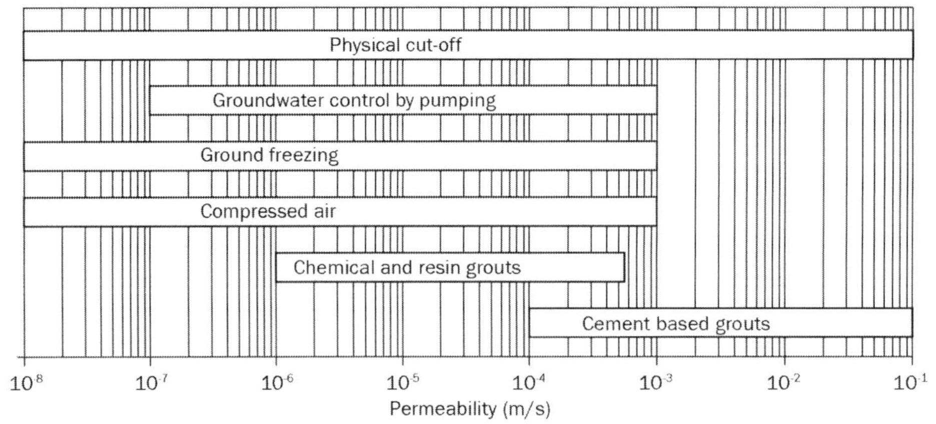


*Figure 1.9 Approximate range of application of groundwater control techniques in soils*

*Table 1.2 Physical cut-off techniques for exclusion of groundwater*

| **Method** | **Typical applications** | **Comments** |
|---|---|---|
| ***Displacement barriers*** | | |
| Steel sheet-piling | Open excavations in most soils, but obstructions such as boulders may impede installation | Temporary or long-term. Rapid installation. Can support the sides of the excavation with suitable propping. Vibration and noise of driving may be unacceptable on some sites, but 'silent' methods are available. See Williams and Waite (1993), and BS EN 12063:1999 |
| Vibrated beam wall | Open excavations in silts and sands. Will not support the soil | A vibrating H-pile is driven into the ground and then removed. As it is removed, grout is injected through nozzles at the toe of the pile to form a thin, low permeability membrane. Relatively cheap. See Privett *et al* (1996) |
| ***Excavated barriers*** | | |
| Slurry trench cut-off wall using bentonite or native clay | Open excavations in silts, sands and gravels up to a permeability of about $5 \times 10^{-3}$ m/s | The slurry trench forms a low permeability curtain wall around the excavation. Quickly installed and relatively cheap, but cost increases rapidly with depth. See Jefferis (1993) |
| Structural concrete diaphragm walls | Side walls of excavations and shafts in most soils and weak rocks | Support the sides of the excavation and often form the sidewalls of the finished construction. Minimum noise and vibration. See Puller and Puller (2003) and BS EN 1538:2010 |
| Secant (interlocking) and contiguous bore piles | As diaphragm walls | As diaphragm walls, but more likely to be economic for temporary works use. Sealing between contiguous piles can be difficult. See Puller and Puller (2003) and BS EN 1536:2010 |
| ***Injection barriers*** | | |
| Jet grouting | Open excavations in most soils and very weak rocks | Typically forms a series of overlapping columns of soil–grout mixture. See Coomber (1986), Rawlings *et al* (2000) and BS EN 12716:2001 |
| Injection grouting using cementitious grouts | Tunnels and shafts in gravels and coarse sands, and fissured rocks | The grout fills the pore spaces, preventing the flow of water through the soil. Equipment is simple and can be used in confined spaces. See Bell (1993), Rawlings *et al* (2000) and BS EN 12715:2000 |
| Injection grouting using chemical and solution (acrylic) grouts | Tunnels and shafts in medium sands (chemical grouts), fine sands and silts (resin grouts) | Materials (chemicals and resin) can be expensive. Silty soils are difficult and treatment may be incomplete, particularly if more permeable laminations or lenses are present. See Bell (1993), Rawlings *et al* (2000) and BS EN 12715:2000 |

continued...

continued from...

| Method | Typical applications | Comments |
|---|---|---|
| *Other types* | | |
| Ground freezing using brine or liquid nitrogen | Tunnels and shafts. Will not work if groundwater flow velocities are excessive (>1 m/day or $10^{-5}$ m/s) | Temporary. A 'wall' of frozen ground (a freezewall) is formed, which can support the side of the shaft as well as excluding groundwater. Plant costs are relatively high. Liquid nitrogen is expensive, but quick; brine is cheaper, but slower. See Harris (1995) |
| Compressed air | Confined chambers such as tunnels, sealed shafts and caissons | Temporary. Increased air pressure (up to 3.5 bar) raises pore water pressure in the soil around the chamber, reducing the hydraulic gradient and limiting groundwater inflow. High running and set-up costs, and potential health hazards to workers. See Slocombe *et al* (2003) |

The various methods of groundwater control using pumped wells, and their main advantages and disadvantages, are summarised in Table 1.3 and described in detail in Chapter 2. Further details of groundwater control methods and applications can be found in Powers *et al* (2007) and Cashman and Preene (2012). Various papers on groundwater control are presented in the proceedings of the 1987 Dublin conference on groundwater effects in geotechnical engineering (see Stroud, 1987) and in a Geological Society publication (Cripps *et al* 1986).

**See also**
Chapter 2 Groundwater control methods

***Table 1.3 Summary of principal pumped well groundwater control methods***

| Method | Typical applications | Comments |
|---|---|---|
| Drainage pipes or ditches (eg French drains) (Section 2.1.2) | Control of surface water and shallow groundwater (including overbleed) | May obstruct construction traffic, and will not control groundwater at depth. Unlikely to be effective in reducing pore water pressures in fine-grained soils |
| Sump pumping (Section 2.1.2) | Shallow excavations in clean coarse soils and in stable rocks | Cheap and simple. May not give sufficient drawdown to prevent seepage from emerging on the cut face of a slope, possibly leading to instability |
| Wellpoints (Sections 2.1.4 and 2.2.2) | Generally shallow, open excavations in sandy gravels down to fine sands and possibly silty sands. Deeper excavations (requiring >5 m to 6 m drawdown) will require multiple stages of wellpoints to be installed | Relatively cheap and flexible. Quick and easy to install in sands. Difficult to install in ground containing cobbles or boulders. Maximum drawdown is ~ 6 m for a single stage in sandy gravels and fine sands, but may only be ~ 4 m in silty sands |
| Deepwells with electric submersible pumps (Section 2.1.5) | Deep excavations in sandy gravels to fine sands and water-bearing fissured rocks | No limit on drawdown. Expensive to install, but fewer wells may be required compared with most other methods. Close control can be exercised over well screen and filter |
| Shallow bored wells with suction pumps (Section 2.1.6) | Shallow excavations in sandy gravels to silty fine sands and water-bearing fissured rocks | Particularly suitable for coarse, high permeability materials where flow rates are likely to be high. Closer control can be exercised over the well filter than with wellpoints |
| Passive relief wells and sand drains (Section 2.1.9) | Relief of pore water pressure in confined aquifers or sand lenses below the floor of the excavation | Cheap and simple. Create a vertical flow path for water into the excavation, water must then be directed to a sump and pumped away |
| Ejector system (Section 2.2.3) | Excavations in silty fine sands, silts or laminated clays in which pore water pressure control is required | In practice drawdowns generally limited to 30 m to 50 m. Low energy efficiency, but this is not a problem if flow rates are low. In sealed wells a vacuum is applied to the soil, promoting drainage |

continued...

continued from...

| Method | Typical applications | Comments |
|---|---|---|
| Deepwells with electric submersible pumps and vacuum<br>(Section 2.2.4) | Deep excavations in silty fine sands, where drainage from the soil into the well may be slow | No limit on drawdown. More expensive than ordinary deep wells because of the separate vacuum system. Number of wells may be dictated by the requirement to achieve an adequate drawdown between wells, rather than the flow rate, and an ejector system may be more economical |
| Electro-osmosis<br>(Section 2.2.5) | Very low permeability soils, eg clays | Only generally used for pore water pressure control when considered as an alternative to ground freezing. Installation and running costs are comparatively high |

The suitability of any of the methods outlined in Table 1.3 depends primarily on the soil permeability, the required drawdown and (if more than one method is technically feasible) the cost. Practical limits to the range of application of each method, in terms of the soil permeability and the drawdown required, are given in Figure 1.10. If the required drawdown and the assessed soil permeability are known, then, by finding the corresponding point on Figure 1.10, an initial assessment can be made of the appropriate groundwater control technique. The shaded areas indicate zones where more than one technique may be suitable.

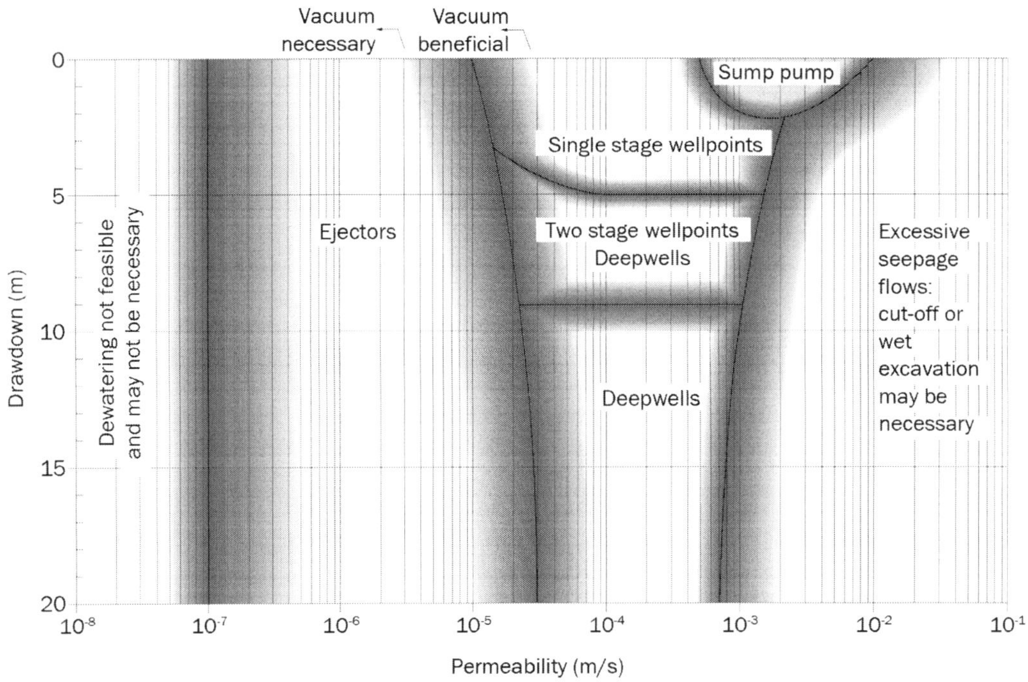


*Figure 1.10 Range of application of pumped well groundwater control techniques in soil (adapted from Roberts and Preene, 1994a, modified after Cashman, 1994b)*

## 1.2.8 Dewatering costs

Relative costs for groundwater control methods using pumping are site specific and depend on ground conditions as well as the method used. Typical cost build ups might include:

- design, planning and environmental permitting costs
- mobilisation and demobilisation of equipment
- installation costs for sumps, wellpoints deepwells, ejector wells or other dewatering installations (for labour, installation equipment and materials such as well screen, filter gravels etc)
- equipment hire costs (for pumps, pipework, electrical controls and monitoring systems)
- fuel and electrical power costs to run pumps
- supervision and monitoring during installation and running (including costs for the designer

to periodically review monitoring data to ensure the system is performing adequately)

- maintenance of plant and rehabilitation of wells if biofouling occurs
- operatives to fuel and maintain pumps
- any charges related to disposal of the discharged water
- backfilling of wells on completion.

## 1.3 KEY REFERENCES


BRASSINGTON, R (2006) *Field hydrogeology, thid edition*, Wiley-Blackwell, London (ISBN: 978-0-470-01828-6)

CASHMAN, P M and PREENE, M (2012) *Groundwater lowering in construction: a practical guide to dewatering, second edition*, CRC Press, Boca Raton, USA (ISBN: 978-0-41921-110-5)

CRIPPS, J C, BELL, F G and CULSHAW, M G (eds) (1986) *Groundwater in engineering geology conference proceedings*, The Geological Society, London (ISBN: 978-0-90331-735-1)

FETTER, C W (2014) *Applied hydrogeology, fourth edition*, Pearson New International Edition, Pearson Education Limited, Essex (ISBN: 978-1-29202-290-1)

POWERS, J P, CORWIN, A B, SCHMALL, P C and KAECK, W E (2007) *Construction dewatering and groundwater control: new methods and applications, third edition*, Wiley-Blackwell, New York, USA (ISBN: 978-0-47147-943-7)

PRICE, M (1996) *Introducing groundwater*, Taylor and Francis, Abingdon, Oxon (ISBN: 978-0-74874-371-1)

STROUD, M A (1987) "Groundwater control – general report". In: *Proc of the 9th conf on soil mechanics and foundation engineering, Dublin, Ireland*, 31 August 1987. E T Hanrahan, T L L Orr, T F Widdis (eds) *Groundwater effects in geotechnical engineering*, vol 1–3, Balkema, Rotterdam, pp 983–1008

YOUNGER, P L (2007) *Groundwater in the environment: an introduction*, Blackwell Publishing, Oxford, UK (ISBN: 978-1-40512-143-9)

# 2 Surface and groundwater control methods

## 2.1 GROUNDWATER LOWERING SYSTEMS

**See also**
Table 1.3 Groundwater control methods

The dewatering systems used today (Table 1.3) have been optimised over many decades of use, although the basic concepts have changed little over the years. Improvements have mainly been in cost reduction from use of new materials, more efficient pumping systems, and faster or more effective installation methods. The physical limitations of the methods have not altered significantly and are unlikely to be improved substantially in the future. The principal systems are described in the following sections.

### 2.1.1 Surface water control

Surface water is not groundwater as such but precipitation and runoff. In free-draining soils of medium to high permeability the surface water tends to drain into the soil down to the groundwater and may be picked up by any dewatering system in operation. In excavations in fine-grained soils, such as sands, silts and clays, of medium to low permeability, surface water might not drain, or only very slowly. In these conditions effective control of surface water is important to prevent batter erosion and softening of the base of the excavation, which would worsen with trafficking of construction plant.

It is good practice to install an effective surface water control system when carrying out an excavation; the need for surface water control may not be obvious when an excavation is first opened, but without it construction plant may become bogged down and work would have to stop after a shower of rain. Surface water can be controlled using systems of drainage blankets, ditches, French drains and garland drains (see Box 2.1). These collect the water and transmit it, usually, to a sump for pumping away (see Section 2.1.2).

### 2.1.2 Sump pumping

**See also**
1.2.5 Instability
4.1 Environmental impacts

Under favourable conditions sump pumping systems can be a simple and cost-effective means of controlling groundwater inflows to an excavation in both soils and rocks. Under unfavourable conditions a sump pumping approach can result in delays, cost overruns and, occasionally, catastrophic failure. The primary limitation on sump pumping is the instability of the soil under the action of the seepage forces generated by the groundwater entering the excavation. This is commonly referred to as 'running sand conditions' or 'boiling' (see Section 1.2.5) and can cause rapid loss of base and side slope stability, leading to a risk of undermining and settlement to adjacent structures. There are too many variables to set simple criteria for when sump pumping is appropriate. The relevant factors to be considered together with favourable and unfavourable conditions for sump pumping are summarised in Table 2.1. The factors in the table are cumulative, so one or two unfavourable conditions may not rule out the use of sump pumping. However, in particular circumstances some factors will be more significant than others. For example, if the works involve heavy foundation loads below the water table in uniform sand, sump pumping is unlikely to be an option, even if all other factors are favourable. If most or all of the factors are unfavourable, it is unlikely that sump pumping would be a viable option.

An important secondary problem with sump pumping is water quality and disposal. Clay, silt and fine sand particles can readily become entrained in the seepage flow, particularly during excavation, and it is virtually impossible to exclude these suspended solids by screening around the sump. The seepage flow may also be susceptible to contamination by cement or any diesel or oil spills from construction plant. Discharge of water contaminated with suspended solids, cement and fuel oils to surface waters, or via soakaway to groundwater, could cause pollution, resulting in environmental damage and the possibility of prosecution by the regulatory authorities. Effective treatment before discharge can prove difficult and costly, but is essential in many cases. These matters are considered further in Section 4.1.

*Table 2.1 Favourable and unfavourable conditions for sump pumping*

| Aspect | Favourable | Unfavourable |
|---|---|---|
| Soil characteristics | Well-graded sandy gravel<br>Clean gravel (expect high flows)<br>Hard fissured rock<br>Firm to stiff clays | Uniform sands and silty sands<br>Soft silts or clays<br>Rock strata with the potential for erosion, softening or instability<br>Sandstone with uncemented layers |
| Hydrology | Modest drawdown<br>No immediate source of recharge<br>Unconfined aquifer | Large drawdown<br>Nearby recharge source<br>Confined aquifer |
| Excavation support | Shallow slopes<br>Deep driven sheet-piling<br>Deep diaphragm wall | Steep slopes<br>Trench sheets with little toe-in<br>Soldier piles and lagging |
| Excavation method | Backactor<br>Dragline | Face shovels<br>Scrapers |
| Structure | Light foundation loads<br>Piled foundations | Heavy foundation loads<br>Pad or strip foundations |
| Environmental requirements | Minimal restrictions on discharge water quality<br>Low risk of contamination of discharge water | Stringent restrictions on discharge water quality<br>High risk of contamination of discharge water |

Sump pumping operations require a system of drains (Box 2.1) to collect the groundwater inflow which, ideally, should be intercepted as it enters the excavation. The drainage system should be sized to deal with groundwater seepage flows and surface water inflows from precipitation. The drainage system should be laid out to feed to one or more sumps, usually located in the corner of the excavation at the deepest point. In large excavations, ditches and French drains should be laid to a fall towards the sump.

*Box 2.1 Water collection methods for surface water control and sump pumping*

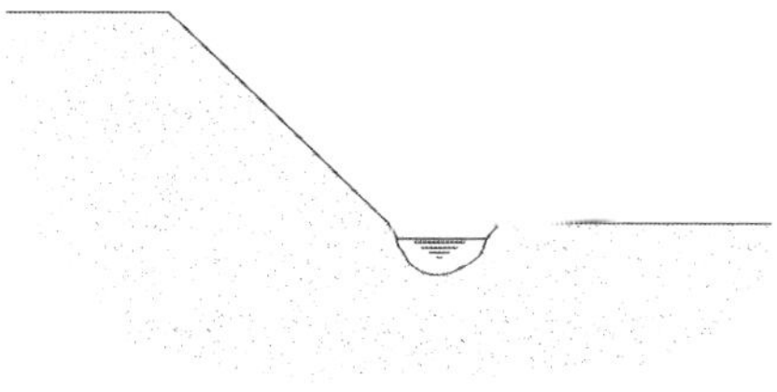

Ditch

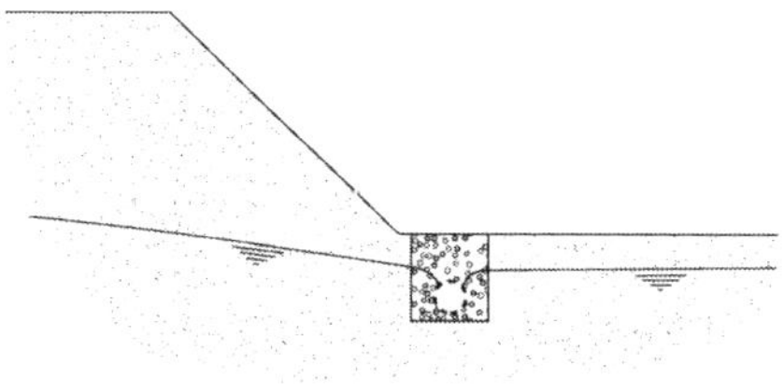

French drain

**Ditch:** these are usually only a viable option in stable ground such as rock or stiff clay. Occasionally a lining is used to control erosion.

**French drain:** consists of a gravel-filled trench typically 0.5 m wide by 0.5 m (or more) deep with a perforated pipe to collect and transmit the flow. Lining the trench with a geotextile filter membrane before placing the gravel and pipe is a useful method for controlling migration of fine soil particles.

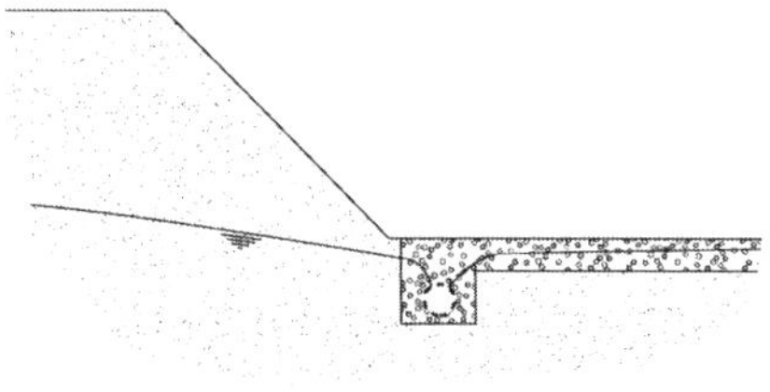

Drainage blanket

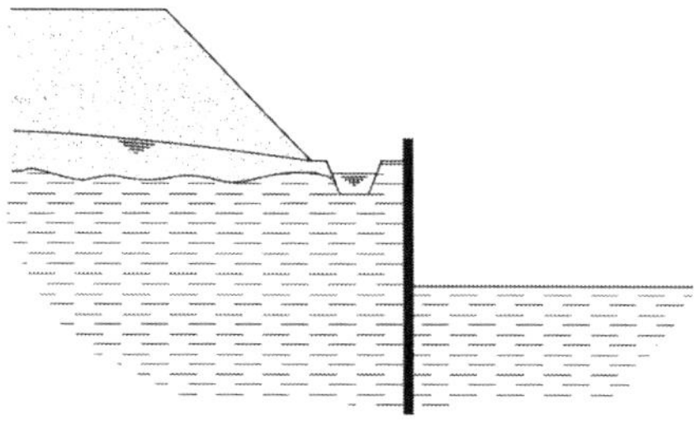

Garland drain

**Drainage blanket:** consists of a layer, 150 mm to 300 mm thick, of free-draining material such as gravel laid on the base of an excavation to collect vertical seepage. The use of a geotextile filter membrane below the drainage blanket is a useful method for controlling migration of fine soil particles. For large areas a network or herringbone of perforated drainage pipes may be needed to transmit the flow.

**Garland drains:** where water enters an excavation as overbleed above an impermeable layer, a garland drain can be used above the base of the excavation to intercept this inflow. Depending on circumstances and soil conditions, garland drains may be channels, ditches or French drains.

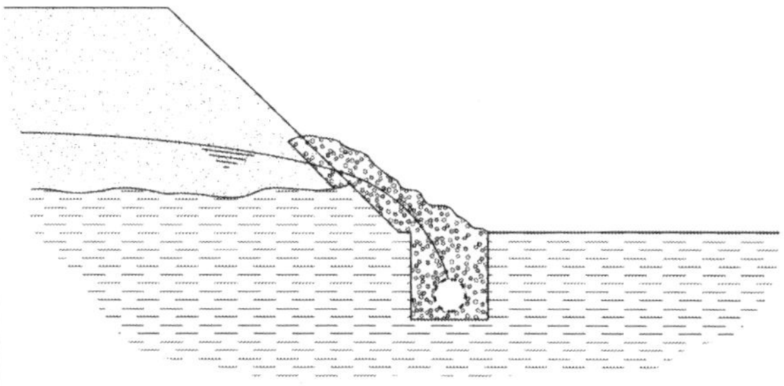

Batter protection

**Batter protection:** where there is a risk of seepage flows emerging on an excavation slope, protection is required to prevent erosion or slope failure. This can be provided by a gravel berm or sandbags.

The requirements for a sump are:

- **Depth:** the sump should be deep enough to drain the excavation and drainage network, allowing for the pump intake level and some accumulation of sediment.
- **Size:** the sump should be substantially larger than the size of the pump to allow space for sediment and cleaning.
- **Filter:** the sump should be perforated or slotted, typically with a hole size or slot width of 10 mm to 15 mm; the sump should be surrounded with coarse gravel (20 mm to 40 mm).
- **Access:** good access is required to allow removal of the pumps for maintenance and cleaning of the sumps to remove any accumulation of sediment.

When excavating it is often necessary to form temporary sumps to control groundwater levels so that a main sump can be constructed for longer-term use. Typical sump arrangements are shown in Figure 2.1.

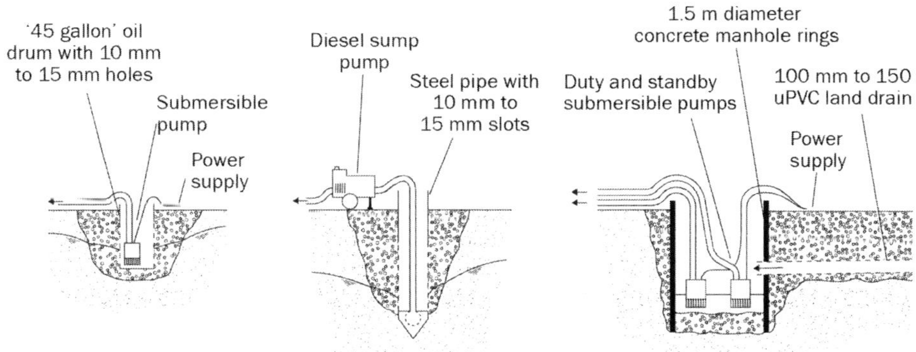


*a) Perforated oil drum* *b) Perforated steel pipe with driving point* *c) Concrete manhole rings fed by French drains*

*Figure 2.1 Typical sumps*

Most sump pumping is carried out using either diesel suction pumps or electric submersible pumps. Pumps are typically available with discharge outlet sizes of 50 mm to 250 mm and with discharge heads of more than 50 m. Diesel suction pumps require no external power supply and sumps can be small because they need only accommodate the suction pipe and strainer. However, suction pumps have a limited lift of approximately 7 m. The question of suction lift does not arise with submersible pumps, but they do require an external power supply and a sump big enough to accommodate them. Hybrid pumps are available, for example hydraulic submersible pumps driven by a diesel hydraulic power pack mounted at the surface. These provide the high discharge head of a submersible pump without the need for an electrical power supply. Typical capacities of sump pumps are given in Table 2.2.

Sump pumping may be used safely for trench excavations in highly permeable soils such as gravel and moderately permeable soils such as sand and gravel mixtures. For drawdowns of more than 1 m to 2 m, inflows can become excessive and unstable conditions may develop; close sheeting will be required to provide trench support. Interlocking trench sheeting can be driven to lengthen drainage paths to limit inflows and control boiling. Where gravel bedding is laid in the base of the trench, this can provide a preferential path for groundwater flows feeding into the excavation area. This problem may also occur where new works are being installed close to existing services laid on gravel bedding (Figure 2.2). The use of clay dams at intervals can limit this transmission of groundwater during construction and in the longer term. Further advice on trench works is given by Irvine and Smith (2001).

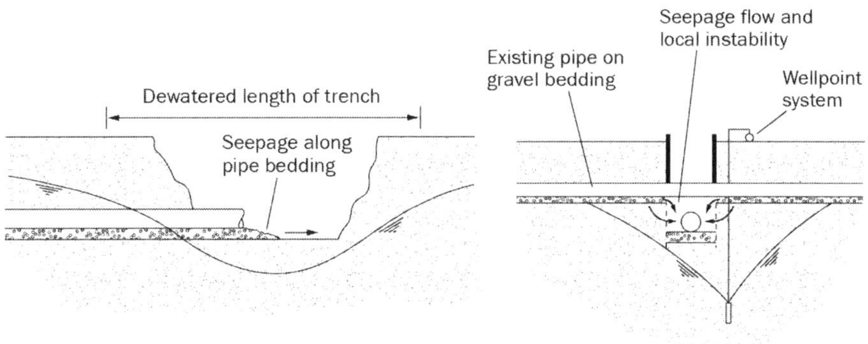


*a) Seepage flow in bedding during construction* *b) Seepage flow along bedding of existing services*

*Figure 2.2 Groundwater flow in pipe bedding*

**See also**
Box 3.3 Settlement tank
4.1 Environmental impacts

When carrying out sump pumping operations, some of the sand and fines fraction in the soil will initially be removed in the immediate vicinity of the sump and drainage network. It is good practice to pass the discharge water through a settlement tank (Box 3.3) to allow the situation to be monitored and to remove those solids that settle readily prior to discharge (see Chapter 4). Settlement ponds or lagoons may be needed to remove any silt or clay fraction present to meet discharge consent requirements (see Sections 4.1 and 4.2). If persistent movement of fines occurs, leading to ground loss and settlement, or if an excavation shows signs of instability, sump pumping should be stopped and supplementary or alternative methods adopted. If the ground loss or instability is serious, it may be necessary to flood the excavation to maintain stability while the situation is reassessed.

## 2.1.3 Wellpoints

Wellpoint systems provide a versatile method of controlling groundwater in a wide range of soil conditions and excavation geometry. A typical wellpoint system layout highlighting the main components is shown in Figure 2.3. Attributes of the wellpoint system are:

**Advantages:**

- flexibility – the same equipment can be used around small and large excavations
- quick to install in many soil conditions
- close spacing (1.5 m to 2 m typically) promotes effective drawdowns in stratified soils.

**Limitations:**

- suction lift of 5 m to 6 m in sands and gravels, but may be limited to 3.5 m to 4.5 m in fine-grained soils
- headermain can cause access restrictions on site.

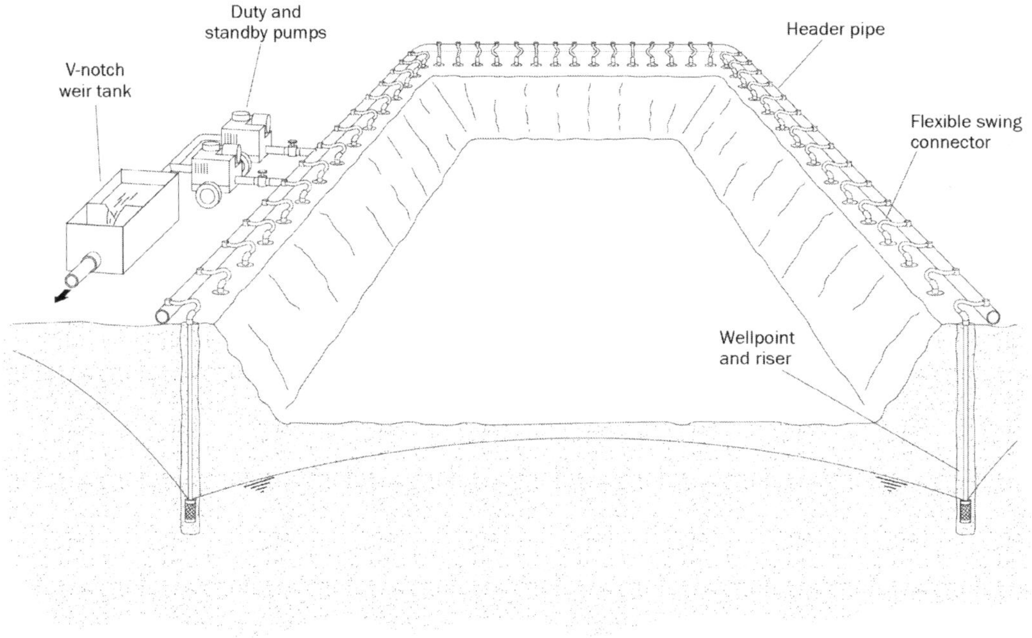


*Figure 2.3 Wellpoint system components*

Wellpoints are essentially shallow wells comprising screens of approximately 50 mm in diameter and 0.5 m to 1 m long. The screens are fitted to the end of a riser pipe typically of 38 mm bore and 5 m to 6 m long. At the surface the riser pipe is linked to the headermain with a flexible pipe referred to as a 'swing'. The swing usually incorporates a valve to allow an individual wellpoint to be turned off or trimmed down if it is drawing air. Headermains are commonly 150 mm diameter pipes, but 100 mm and 200 mm equipment is also available. The headermain connects to a vacuum pump capable of handling large volumes of both air and water. The pumps are generally vacuum-assisted self-priming centrifugal pumps driven by diesel or electric motor. Positive-displacement piston pumps are also available and can be very economical in power consumption where flows are modest. Typical capacities of pumps are given in Table 2.2.

*Table 2.2 Examples of sump pump and wellpoint pump capacities*

| Pump type | Power (kW) | Discharge outlet size (mm) | Working head (m) | Flow (l/s) |
|---|---|---|---|---|
| Sump pump: Electric submersible | 4.6 | 75–100 | 10<br>15 | 18<br>11 |
| | 9.5 | 100–150 | 10<br>20 | 45<br>22 |
| | 23 | 150 | 10<br>25 | 85<br>50 |
| | 41 | 200 | 10<br>25 | 180<br>100 |
| Sump pump: Rotary suction self-priming | 5.5 | 100 | 10<br>15 | 30<br>20 |
| | 11 | 100–150 | 10<br>15 | 45<br>35 |
| | 15 | 150 | 10<br>15 | 60<br>45 |
| | 22 | 200 | 10<br>15 | 100<br>70 |
| Wellpoint pump: Rotary suction plus exhauster for air | 15 | 100–150 | 10<br>15 | 40<br>25 |
| | 22 | 150 | 10<br>15 | 55<br>35 |
| Wellpoint pump: Piston suction (positive displacement) | 5.5 | 100 | 10<br>15 | 18<br>18 |
| | 7.5 | 125 | 10<br>15 | 26<br>26 |

**Note**

Working head is the suction head plus the discharge head and friction losses

## Wellpoint spacing

For a particular project the number of wellpoints required and their spacing depends on several factors:

- permeability of the soil and expected seepage flows
- soil stratification and risk of overbleed flows
- excavation geometry and perimeter length
- required drawdown.

Typical spacings for a range of conditions are shown in Table 2.3.

*Table 2.3 Typical wellpoint spacing*

| Permeability | Uniform soil conditions | Stratified soil or overbleed risk |
|---|---|---|
| High (>$10^{-3}$ m/s) | 1.0–1.5 m | 1.0–1.5 m |
| Medium ($10^{-3}$–$10^{-5}$ m/s) | 1.5–3.0 m | 1.0–2.0 m |
| Low (<$10^{-5}$ m/s) | 1.5–2.0 m | 1.0–2.0 m |

The maximum capacity of a standard 50 mm diameter wellpoint with a screen length of 0.75 m and a 0.5 mm filter mesh is approximately 1 l/s in high permeability soils. In such soils the spacing of the wellpoints is dictated by the perimeter length of the excavation and the flow capacity required to achieve drawdown. If the wellpoint spacing needs to be less than about 1 m, wellpoint dewatering may not be the most appropriate technique for the works. In certain applications yields can be increased by using larger-diameter high-capacity wellpoints or by installing two or more wellpoints in one hole. Alternative options might be sump pumping (Section 2.1.2), high-capacity suction wells (Section 2.1.6), or physical exclusion of the groundwater with cut-offs (see Table 1.2).

In homogeneous soils of medium permeability individual wellpoint yields are limited by the soil permeability, k, and wellpoint spacings of 1.5 m to 2 m are typical. It is sometimes possible to extend the wellpoint spacing to 3 m or more if shallow drawdowns, ie 3 m or less, are required in soils where the permeability is in the middle of the range of Table 2.3 (around $k = 1 \times 10^{-4}$ m/s).

For stratified soils containing layers or pockets of silt and clay, a close wellpoint spacing is recommended for effective drainage of all layers, particularly where drawdown to an impermeable layer is required. Spacings of about 1.5 m are typically used in this situation. Even with a close wellpoint spacing, it is not possible to achieve full drawdown to an impermeable interface; some overbleed inflow into the excavation is unavoidable. Control measures (possibly using sandbags or a gravel berm to provide slope stability in fine-grained soils together with a perimeter drain) and sump pumping may be necessary (Figure 2.4). If soil conditions permit, wellpoints can be 'toed in' to the underlying impermeable stratum to create a local sump. Where this is not feasible, short-screen wellpoints, 300 mm to 400 mm long, can be used to maximise drawdowns.

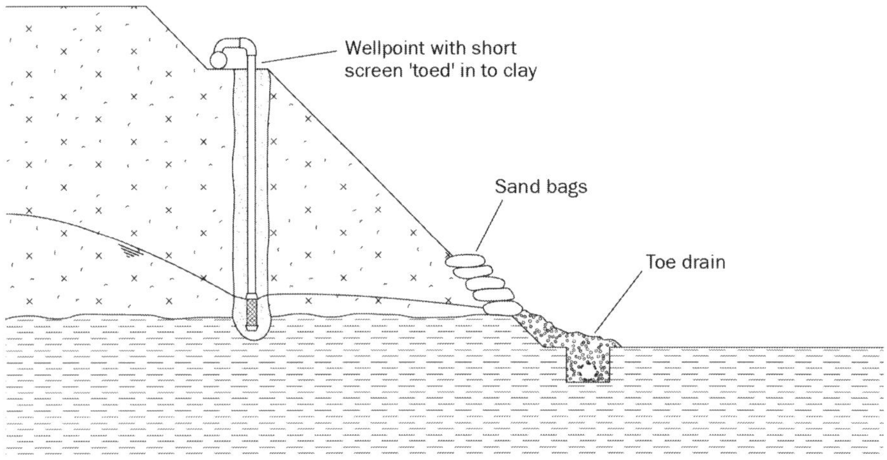


*Figure 2.4 Control of overbleed seepage flows*

## Suction lift

**See also**
2.2.2 Vacuum wellpoints

The main limitation on the performance of wellpoint schemes is suction lift. Although the maximum lift at sea level is theoretically just over 10 m, in practice efficiency losses reduce this to about 6 m at the wellpoints. If a wellpoint system is installed above sea level, the suction lift will be further reduced because of the lower atmospheric pressure. For every 300 m elevation above sea level, the maximum suction lift of a wellpoint system is reduced by about 0.3 m. Furthermore, in fine-grained soils of medium to low permeability some suction may be needed to induce drainage, so the suction lift could be reduced to approximately 3.5 m to 4.5 m (see Section 2.2.2).

Where drawdowns of more than 5 m are required, multi-stage wellpoint systems can be used, as shown in Figure 2.5. Under favourable conditions successive wellpoint stages can be placed at about 4.5 m depth intervals but the lower stages take up space within the excavation. Pumping on lower stages often diverts water from the upper stages, allowing pumping of these to be discontinued.

## Wellpoint installation

Wellpoints are usually installed by jetting. Plastic disposable wellpoints are most commonly used, but the older style steel self-jetting reusable wellpoints remain available and can prove useful for particular applications, eg where headroom or access is restricted. Typical examples of both types of wellpoint are shown in Figure 2.6. The techniques used for wellpoint installation are summarised in Table 2.4.

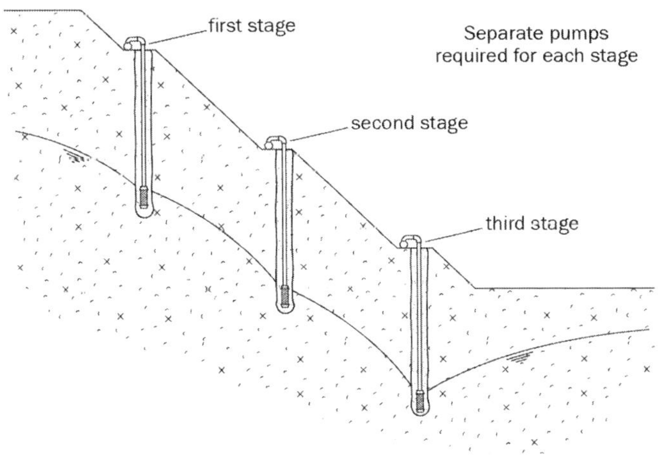


*Figure 2.5 Multi-stage wellpoint system*

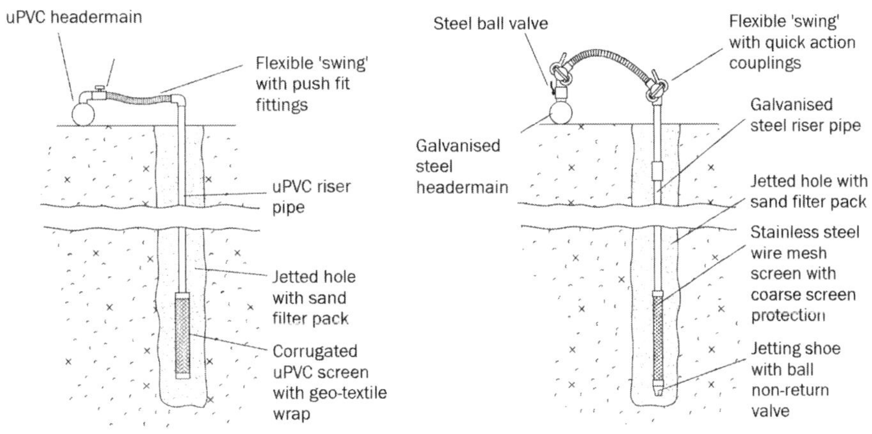


*Figure 2.6 Disposable and reusable wellpoints*

Figure 2.7 shows the installation of steel self-jetting wellpoints. The steel riser pipe is sufficiently rigid to allow water to be fed to the top of the 6 m long riser pipe from a jetting pump. The jet of water from the cutting shoe allows rapid penetration in sandy soils down to about 5 m or 6 m in a few minutes. Usually, filter sand is introduced into the jetted hole once the wellpoint has been installed to depth. This is a skilled operation because the introduction of the sand has to be co-ordinated with shutting off the jetting pump to achieve effective sand placement. On completion of the dewatering works the wellpoints can be pulled out with an excavator or crane for reuse.

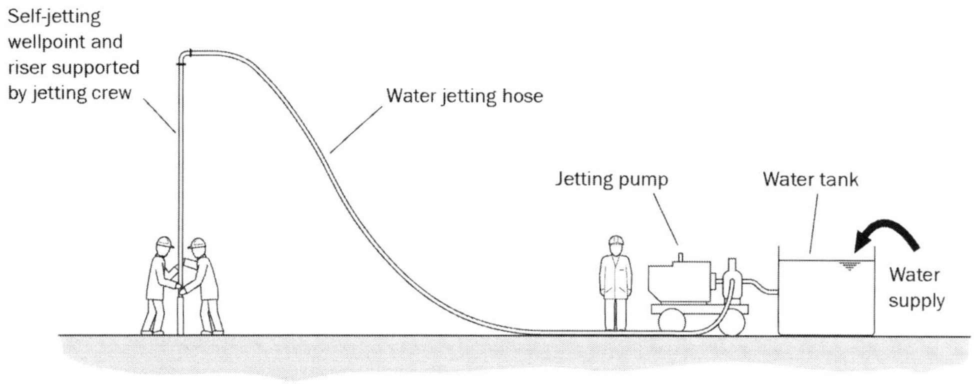


*Figure 2.7 Installation of reusable steel self-jetting wellpoints*

*Table 2.4 Summary of principal wellpoint installation techniques*

| Method | Resources | Typical diameter and depth of bore | Notes |
|---|---|---|---|
| Self-jetting wellpoint (Figure 2.7) | Supervisor<br>Two labourers<br>Jetting pump | 100 mm uncased to approx. 7 m depth | Not widely used<br>Useful if access is restricted<br>Effective in non-cohesive silt, sand and sandy gravel |
| Placing tube (Figure 2.8) | Supervisor<br>Labourer<br>Excavator operator<br>Placing tube<br>Jetting pump<br>(Compressor)<br>Excavator or crane | 100 mm to 150 mm cased to approx. 10 m depth | Most commonly used system for disposable wellpoints<br>Effective in non-cohesive silt, sand and sandy gravel |
| Auger pre-drilling (Figure 2.9) | Supervisor<br>Excavator operator<br>Hydraulic auger unit<br>Excavator | 150 mm to 300 mm uncased to approx. 7 m depth | Used for pre-drilling superficial cohesive strata before installation with placing tube |
| Cable percussion drilling (Figure 2.10) | Supervisor<br>Drill rig operator<br>Assistant driller<br>Cable percussion rig and casing | 150 mm to 300 mm cased<br>30 m depth and more | Effective but slow<br>Can penetrate a wide range of cohesive and non-cohesive soils and weak rock |
| Rotary jet drilling (Figure 2.11) | Supervisor<br>Labourer<br>Drill rig operator<br>Jetting pump<br>(compressor)<br>Drill rig | 100 mm to 250 mm cased<br>15 m depth and more | Rapid installation rates possible<br>Effective at penetrating clays, silts, sands, sandy gravels and weak rock |

Plastic disposable wellpoints are installed by jetting using a temporary steel placing tube (Figure 2.8). The wellpoint is then installed and any filter sand is introduced to the jetted hole as the temporary steel casing is withdrawn.

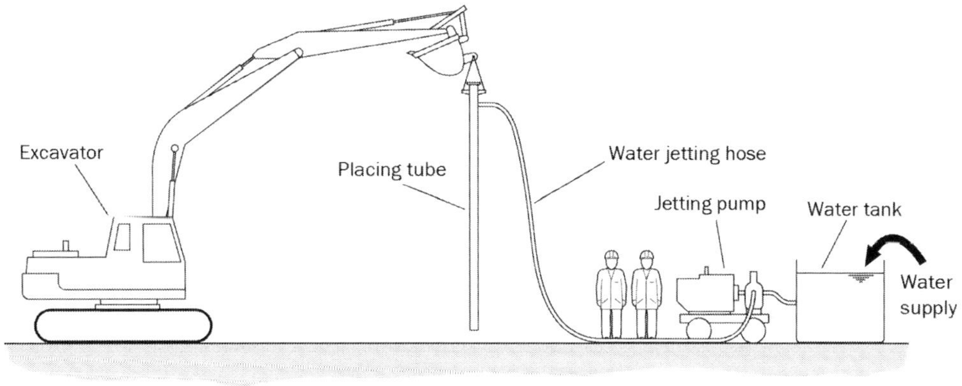


*Figure 2.8 Wellpoint installation by placing tube*

**See also**
Chapter 4 Environmental matters

The jetting water runoff can lead to rapid deterioration of surface conditions on some sites. Moreover, unintentional discharge into surface waters could cause pollution resulting in environmental damage and the possibility of prosecution by the regulatory authorities (see Chapter 4). In order to avoid this it is good practice to excavate a shallow trench, say 0.5 m wide by 0.5 m deep, along the line of the proposed wellpoint system before jetting to contain the runoff. If a sump is being used to provide the supply of jetting water, it is sometimes possible to recirculate the water by channelling it back to the sump.

In sands and very sandy gravels installation by jetting is an effective and economical method. However, it can prove difficult to jet through clay or clayey soils to dewater a more permeable underlying stratum; pre-augering a hole through the clay using an excavator-mounted auger can be very effective (Figure 2.9).

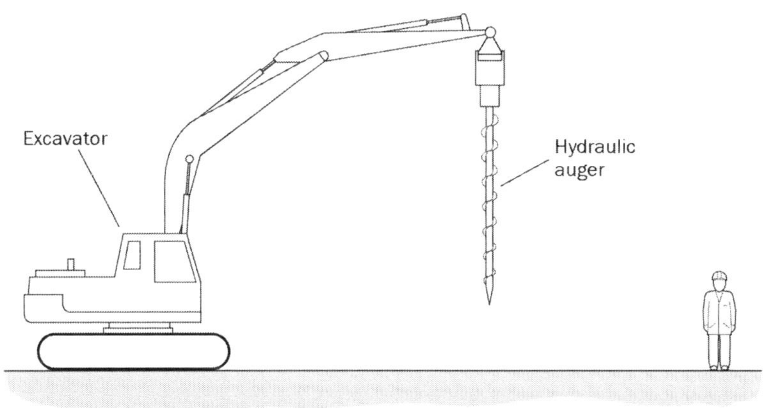


*Figure 2.9 Excavator-mounted auger for pre-drilling of clays*

Installation of wellpoints by cable percussion drilling is occasionally undertaken (Figure 2.10) usually where there is a need to control drilling spoil and returns or to ensure effective installation of annular seals. This may be a requirement where there is concern that contamination may be present particularly where more than one water-bearing horizon is evident. A significant benefit of cable percussion drilling is that the soil profile can be established by logging soil samples recovered from the bore prior to completing the wellpoint installation. This is in contrast to jetting techniques, which provide minimal information on the soil profile.

Soils in which 'loss of boil' occurs usually have a permeability at or close to the upper limit for effective wellpoint dewatering. Such installation difficulties could be an early indication of future problems, with very high flow rates making the required drawdown difficult to achieve.

Rotary jet drilling (Figure 2.11) can be a cost-effective method of wellpoint installation. A drill rig with a hydraulic head and swivel allows a temporary open-ended steel casing to be rotated as it is jetted into the ground. This system is versatile and can achieve fast installation rates through a range of conditions including clays, sands, sandy gravels and weak rock.

*Figure 2.10 Wellpoint installation by cable percussion drilling (courtesy WJ Groundwater Limited)*

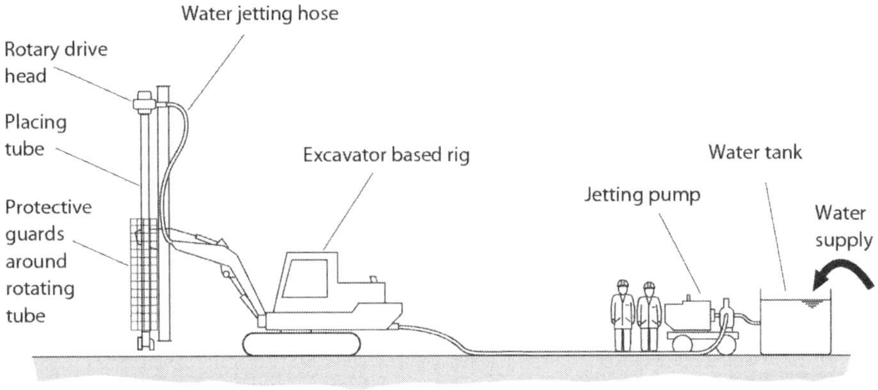


*Figure 2.11 Wellpoint installation by rotary jet drilling*

## Use of filter sands in wellpoint installations

In appropriate conditions, a column of filter sand (known as a filter pack) is introduced around each wellpoint during installation as shown in Figure 2.6. The purpose of this filter pack is both to provide a vertical drainage path around the wellpoint and to allow the wellpoint screen to be matched to the grading of the soil.

**See also**
6.3.3 Filter design

The provision of a vertical drainage path is an important requirement where there are stratified soils and perched water to be drained. In coarse well-graded soils, such as sandy gravel where $D_{40} > 0.5$ mm, it is not generally necessary to install a filter pack around a wellpoint. This is because an effective natural filter pack can be developed by careful control of the jetting water after the wellpoint has been installed. In these conditions there is little risk of persistent pumping of fines or clogging of wellpoint screens. However, in fine-grained poorly graded soils, such as uniform fine sand, a filter pack is essential to maximise wellpoint performance and avoid persistent pumping of fines. Appropriate filter material for wellpoint installation is typically medium to coarse sand, such as a sharp concreting sand. For particularly difficult conditions and further information on this topic see Section 6.3.3.

## Trimming of wellpoint systems

As the water table is lowered, some wellpoints may begin to draw in air, causing a loss of vacuum. If excessive, this can prevent the required drawdown being achieved. In order to avoid this, the flow from each wellpoint should be controlled using the valve on the swing connectors linked to the headermain. Each valve is adjusted or throttled back until the flow is smooth and then re-opened slightly. This procedure is termed 'trimming' or 'tuning' of the wellpoint system. The process is iterative; trimming of one wellpoint will affect others in the system. If the soil stratification allows, trimming can be reduced by installing wellpoints with 9 m long riser pipes. The suction limitations of a wellpoint system mean that air cannot be readily drawn into such a system.

## Wellpoint system layout for open excavations

Wellpoint systems are typically installed in a ring configuration around an excavation, as illustrated in Figure 2.3. It is generally beneficial to carry out an initial excavation to within about 0.5 m of the standing groundwater level before deploying the wellpoint system. This facilitates the wellpoint installation, saving time, and, provided the pumps and headermain are installed at the lower level, reduces the required lift and maximises system performance.

A typical 150 mm wellpoint dewatering pump is capable of pumping 50 to 100 individual wellpoints. It is advisable to provide standby pumps to cover for mechanical failure or stoppage of the duty pumps. Standby pumps should be plumbed into the headermain and discharge pipes so that they are ready for immediate use in an emergency. The headermain and pumps should be maintained at the same approximate level for optimum performance. This may create access restrictions to an open excavation, which can be overcome by either leaving out a number of wellpoints and providing ramps over the headermain, or by leaving a gap in the headermain at the end of the line of wellpoints. Access is also required to individual wellpoint valves for trimming; it is inadvisable to completely cover or bury sections of the wellpoint system except at agreed plant crossings.

## Wellpoint systems in conjunction with sheet-pile cofferdams

Steel sheet-pile cofferdams can be used to provide excavation side support. Where dewatering is required in conjunction with a cofferdam, careful consideration has to be given to the interaction between the flow of groundwater to the dewatering system and the sheet-piles. In particular it is important to understand the pore water pressure regime that will result from the dewatering works and check that the design of the cofferdam is adequate for both the soil loading and the hydrostatic loads that may arise. Some examples are given in Box 2.2. The design and construction procedures for sheet-pile cofferdams are discussed by Williams and Waite (1993), Gaba *et al* (2003), BS EN 1997-1:2004 and in BS EN 12063:1999.

## Wellpoint system layouts for trench works

An important use for wellpoint systems is for trench works dug below the standing groundwater level. The basic layout options are either a single-sided system or a double-sided system, as shown in Figure 2.12. The advantages and disadvantages of these methods are summarised in Table 2.5.

**Example 1 Box culvert with internal wellpoints.**

A box culvert was constructed below the standing groundwater level in storm beach gravels overlying a dense silty fine sand. The invert level for the culvert was in the sand stratum. Excavation side support was provided by a steel sheet-pile cofferdam. Dewatering was carried out initially by sump pumping to allow much of the gravel to be removed, followed by internal wellpoint dewatering (see figure). Removal of much of the gravels was necessary to facilitate wellpoint installation. As the superficial storm beach gravels are highly permeable, no external drawdowns would be developed by the internal system. The cofferdam was designed to take full external hydrostatic loads. The wellpoints had only to deal with the modest flows from the underlying silty fine sand. Dewatering without sheet-piles was not an option because of the very high permeability of the storm beach deposits.

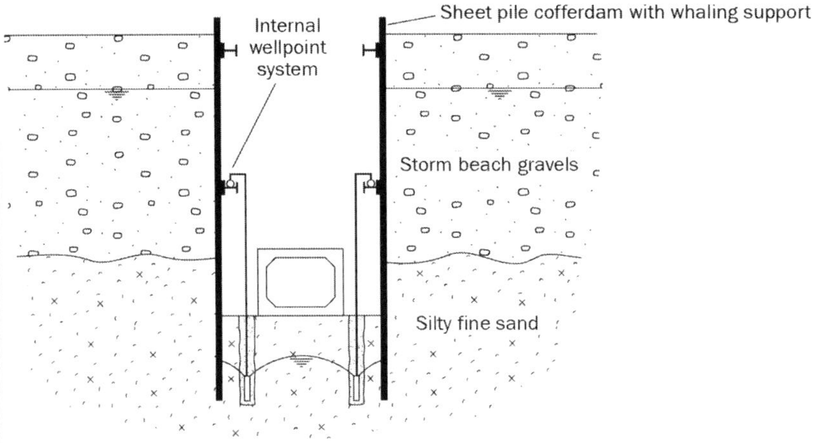


**Example 2 Cantilever sheet pile retaining wall with external wellpoints**

A basement excavation in fine to medium sand used cantilever sheet-piles for side support. Propping of the sheet-piles was problematic because of the width of the excavation (40 m) and because the sheet-piles were to be left in place and used as a back-shutter when casting the basement wall. Dewatering was carried out with an external ring of wellpoints (see figure). The external drawdown removed the hydrostatic loading on the sheet-piles, avoiding the need for other support. Monitoring, maintenance and reliability of the dewatering system was important, because a stoppage in pumping could result in recovery of the groundwater levels and catastrophic failure of the cantilever sheet-piles.

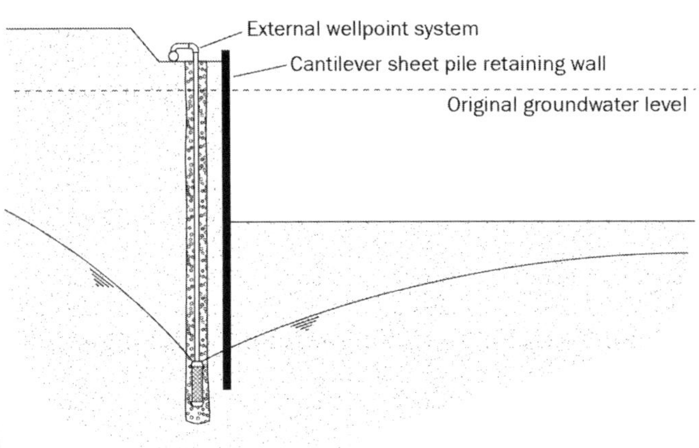

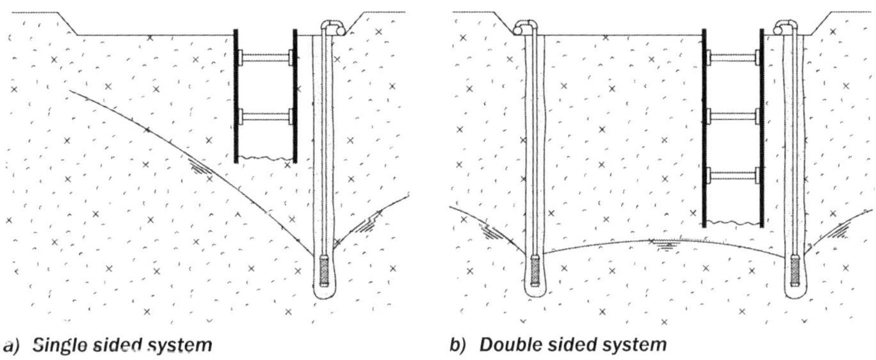


*Figure 2.12 Wellpoint systems for trench works*

*Table 2.5 Advantages and disadvantages of single-sided and double-sided systems for trench works*

| Aspect | Single-sided system | Double-sided system |
|---|---|---|
| Access | Good access maintained on one side | Access restricted on both sides |
| Trench width | Typically limited to about 2 m | Effective for excavations 10 m wide or more |
| Trench depth (below headermain and pump) | Typically limited to about 4.5 m | Limited to 5.5 m for a single-stage system |
| Soil conditions | Not suitable in low permeability soils due to steep cone of drawdown<br>Requires permeable soil to an adequate depth below formation | Effective in a wide range of soil conditions<br>Has to be used in stratified soils or if an impermeable layer is present above or close to formation |

For trench works drawdowns normally need to be developed rapidly and a wide wellpoint spacing is therefore inappropriate. A wellpoint spacing of 1 m to 2 m is typical.

For trenches less than about 120 m long, a static wellpoint system is appropriate, ie wellpoints are installed and connected to the pumping main for the whole length. For trench works longer than about 120 m it may be cost-effective to use a progressive system where disposable wellpoints are installed for the full length of trench, but the headermain and pumps are initially connected for the first length only (typically 60 m to 100 m). These then 'leap-frog' forward as the excavation progresses (see Figure 2.13). Valves in the headermain can allow sections to be isolated and progressed. A sufficient length of wellpoint equipment has to be operational both ahead of and behind the length of open trench to provide effective drawdown.

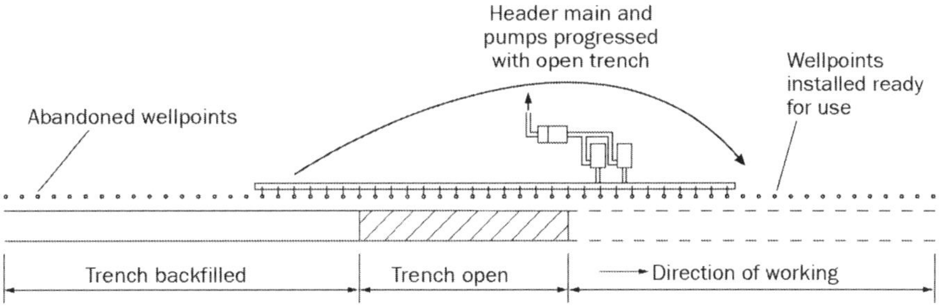


*Figure 2.13 Progressive wellpoint system for trench works*

## 2.1.4 Horizontal wellpoints

The horizontal wellpoint system consists of a perforated pipe (the well screen), which is laid horizontally in the base of a trench. The trench is backfilled and, as illustrated in Figure 2.14, the screen feeds to a non-perforated suction pipe, which is brought to the surface at intervals and is pumped directly by a standard wellpoint vacuum pump. The perforated pipe is normally corrugated PVC of 80 mm to 150 mm diameter wrapped in a geotextile filter mesh. The pipe is typically laid in lengths of up to 100 m at a depth of between 2 m and 6 m. The design principles, including the suction lift limitation of about 6 m, are similar to those for a conventional wellpoint system. In appropriate soil conditions it can be beneficial to install filter sand around the perforated pipe when it is laid.

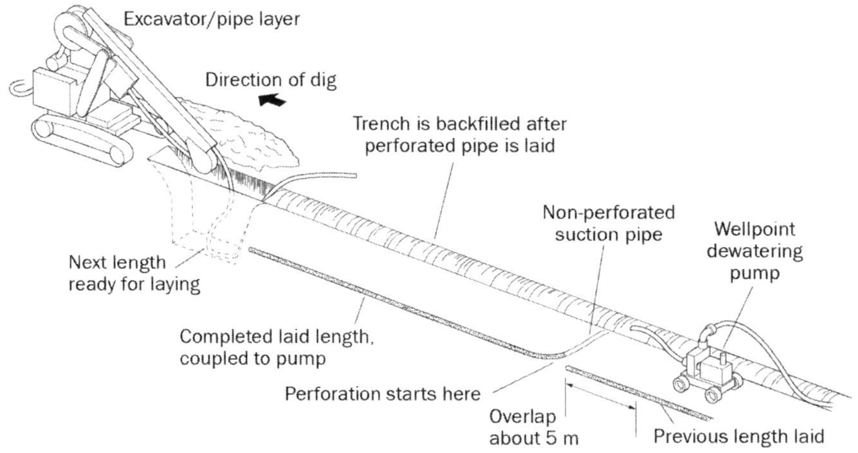


*Figure 2.14 Horizontal wellpoint installation using a land drain trenching machine*

The cost-effectiveness of a horizontal wellpoint scheme depends primarily on the speed and cost of the drain installation. Conventional trench excavation techniques using a backactor can be used, but this is relatively slow. In unstable water-bearing ground a conventional wellpoint system would probably be needed for construction of the trench, which means that such methods are unlikely to be cost-effective. For large-scale use, horizontal wellpoint systems have only proved to be viable using special land drain installation trenching machines (Figure 2.14). Machines are available that can excavate a trench 225 mm wide to a depth of between 2 m and 6 m, lay a flexible perforated pipe and backfill the trench in one continuous operation.

Attributes of the horizontal wellpoint system are:

**Advantages:**

- provides a clear working area without access restrictions at ground level
- with a specialist trenching machine fast installation rates can be achieved (up to 1000 m per day in good conditions)
- particularly suitable for long pipe-laying contracts
- jetting water is not required for installation
- once the drainage pipe is laid, set-up and dismantling is simple and fast.

**Limitations:**

- suction lift is limited to 5 m to 6 m
- specialist trenching machines weigh up to 32 tonnes and are too heavy to be used in soft soils; machines may have to be fitted with wide tracks to reduce ground pressures
- wear and damage to trenching machines can be severe where the ground conditions are coarse gravel or where cobbles and boulders are present
- difficulties can arise if a layer of soft clay is present in the trench and the water table is high; the clay may 'slurry up' and coat the perforated pipe, thereby clogging it as it is laid
- mobilisation and demobilisation costs for large trenching machines are high; this severely restricts their use on small contracts.

Large trenching machines were used relatively widely to install horizontal wellpoint systems for the dewatering of gas pipelines and motorway cuttings in the 1960s and 1970s. Currently trenching machines capable of installation at 6 m depth are not widely available in the UK and as a result horizontal wellpoint schemes are rarely used. Such equipment is more widely used in other countries where conditions are more suited to their capability. Further details of the method are given in Brassington and Preene (2003).

## 2.1.5 Deepwells with submersible pumps

**See also**
5.3.1 Pumping tests
6.3.1 Well losses
Figure 6.14 Well losses

In a deepwell system the suction lift limitation is overcome by placing the pump down the well. Slim-line electric submersible pumps are widely available, being commonly used in water supply wells. With the pump installed near the base of the well, the only limit on drawdown in the well is the power and performance of the submersible pump deployed. The external drawdown that can be achieved by a single well installed in a water-bearing formation is generally not great relative to the depth of the well (see Figure 6.14). This is because of the high losses generated by the concentrating effect of the radial flow in the vicinity of the well (see Section 6.3.1). It is usually necessary to install an array of several deepwells to achieve a desired drawdown over a specified area.

The design of deepwell systems is more complex than for wellpoint systems. This is because deepwell arrays rely on interaction of drawdowns remote from the wells to achieve the desired effect. This 'action at a distance' requirement can make deepwell systems susceptible to local variations in ground conditions. The availability of comprehensive, good quality site investigation data, ideally including a pumping test (see Section 5.3.1), is important for the successful design and specification of deepwell systems. A case study of a deepwell scheme where a very significant local variations in conditions was identified and successfully managed is described in Bevan *et al* (2010).

A typical deepwell system layout is shown in Figure 2.15. Attributes of the deepwell system are:

**Advantages:**

- drawdown only limited by depth of well and soil stratification
- pressure relief can be provided in deep layers
- wells can be placed away from working areas (for example at the top of batters)
- wells are usually installed at relatively wide spacing, which minimises surface access restrictions.

**Limitations:**

- relatively high installation costs per well means the number of wells should be optimised
- comprehensive, good quality site investigation information is required for design
- flexibility of equipment is restricted because individual pumps cover a limited range of flow and discharge head
- pumps are electrically powered so both duty and standby power supplies are required for reliability.

Each deepwell consists of a well liner with submersible pump installed as shown in Figure 2.16. The well liner has a perforated screen section which allows the entry of groundwater.

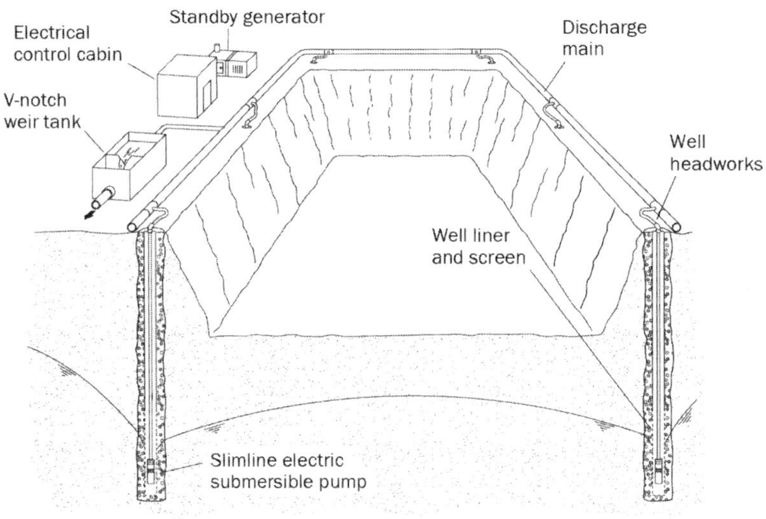


*Figure 2.15 Deepwell system components*

## Pumps and pipework

**See also**

A1 Datasheet 2 Pipework friction losses

The most common deepwell pumps are slim-line multi-stage rotary electric submersible pumps, designed to be of minimum external diameter. Examples of the minimum internal diameter of well liner necessary to accommodate pumps of various capacities are given in Table 2.6. The pump capacities given in Table 2.6 are the maximum, typical operating flows are 10 to 20 per cent lower.

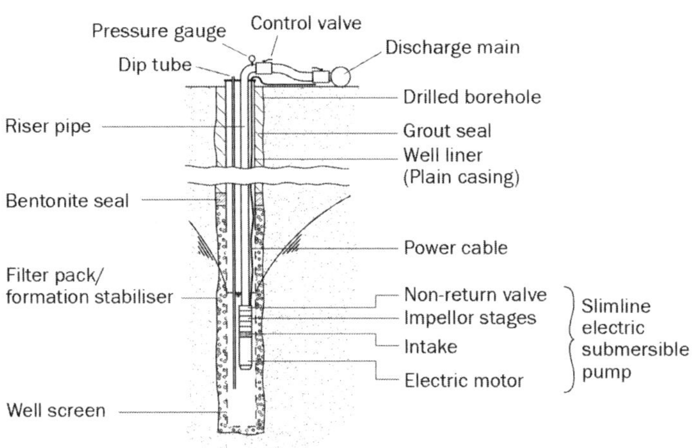


*Figure 2.16 Schematic section through a deepwell*

*Table 2.6 Typical minimum well liner diameters for slim-line submersible borehole pumps*

| Maximum pump capacity l/s | Pump discharge size mm (inches) | Pump diameter mm | Minimum well liner i.d. mm |
|---|---|---|---|
| 2.4* | 38 (1.5) | 74 | 76 |
| 5 | 63 (2.5) | 101 | 110 |
| 10 | 76 (3) | 134 | 145 |
| 20 | 102 (4) | 146 | 152 |
| 30 | 125 (5) | 200 | 203 |
| 45 | 125 (5) | 200–204 | 254 |
| 90 | 152 (6) | 258 | 306 |

**Notes**

* only available with 200 to 230 volt motor.

Includes information from more than one pump manufacturer.

For each pump diameter and capacity there is a family of pumps covering discharge heads from 10 m to 200 m or more. The discharge head is increased by adding stages to the rotary pump. To provide the increased discharge head, electric motors of increased power are required. Borehole pumps are often manufactured entirely from stainless steel, although certain manufacturers incorporate some plastic, cast iron or bronze components. Higher specification pumps are available for use where groundwater is saline, or where severe corrosion conditions are anticipated.

A typical arrangement for the pump and pipework in a well is shown in Figure 2.16. The pipework should be of sufficient size not to incur excessive head losses, which could adversely affect the pump performance. Information on head losses in pipework and valve systems is given in Datasheet 2 (Appendix A1).

## Well liners and screens

The pump unit is installed in a well liner and screen, which should have the following characteristics:

- sufficient internal diameter to accommodate the pump and any electrical control gear (see Table 2.6)
- sufficient strength to support soil loads together with any hydraulic pressures developed during operation without collapse or distortion
- resistance to corrosion in the prevailing geochemical environment
- a screened section capable of retaining the soil and filter pack with the minimum resistance to the groundwater flow entering the well.

**See also**
6.3.3 Filter design

When selecting a well screen, the most important parameter to consider is the aperture size, which should match the grading of the surrounding soil and any annular filter pack. Also of significance is the 'free open area'. This is the total area of the apertures expressed as a percentage of the total screen area. A screen with a larger 'free open area' should give reduced resistance to groundwater inflow, providing it is installed and developed correctly and where necessary has an appropriate filter pack. Design procedures for the specification of a well screen and annular filter pack are covered in detail in Section 6.3.3.

A summary of the most common commercially available well screens is given in Table 2.7. The simple slotted PVC screens are effective in a range of conditions and are widely used. The more sophisticated screens offer either durability or increased free open area or both, at a cost. A comparison between aperture size and free open area for commercially available screens is given in Table 2.8. Selection of slot or geotextile aperture size is discussed in Section 6.3.3.

*Table 2.7 Summary information on commercially available well screens*

| Pipe material | Screen type | Range of sizes o.d. by i.d. mm | Aperture size mm | Notes |
|---|---|---|---|---|
| PVC | Slots | 32 by 28 to 630 by 593 | 0.30–5.0 | Wide range of slot and pipe sizes<br>Readily available |
| Thermoplastic | Slots | 78 by 51 to 350 by 299 | 1.5–5.0 | Strong, durable and inert<br>Difficult to cut fine slots |
| Thermoplastic or PVC (base pipe) | Geotextile<br>3 layer | 78 by 51 to 350 by 299 | 0.10–0.6 | Very fine aperture sizes available |
| Galvanised or stainless steel | V-wire continuous slot | 60 by 39 to 610 by 577 | 0.5–2.0 | Very strong, high quality, durable<br>Only built to order |
| Galvanised or stainless steel | Louvered or punched holes | 105 by 90 to 1015 by 980 | various | Not widely used for dewatering wells |

*Table 2.8 Comparison of typical free open areas for various screen types*

| Aperture size mm | Slotted % | Three-layer geotextile % | Steel V-wire slot % |
|---|---|---|---|
| 0.01 | | 7.5 | |
| 0.15 | | 7.5 | |
| 0.25 | | 20 | 10 |
| 0.30 | 4 | | |
| 0.40 | 5 | 23 | |
| 0.50 | 6 | | 15 |
| 0.60 | 8 | 23 | |
| 1.0 | 11 | | 27 |
| 1.5 | 16 | | 35 |
| 2.0 | 20 | | 42 |
| 3.0 | 25 | | |
| 4.0 | 28 | | |

PVC pipe can be obtained with a wall thickness of just a few millimetres. With the rise in the use of PVC screens, there have been a few cases of well screens collapsing, even though soil loads appear to be well within the collapse resistance of the liner. A number of factors may have contributed to these collapses:

- significant hydraulic loading can be generated across screens by rapid drawdown of the water in a well when pumping; particularly if the screen is too fine or if drilling mud remains outside the screen
- heat generated by grout curing can cause softening of plastic well liners
- the collapse resistance of slotted screen and joints is lower than for plain casing
- pile installation by vibrator or drop hammer can cause excess loading to nearby wells because of local soil liquefaction
- liner and screens can be damaged by mishandling during installation.

## Well construction

**See also**
Chapter 4 Environmental matters
6.3.2 Well depth
6.3.3 Filter design

Wells are constructed by boring a hole, usually by cable percussion drilling, rotary drilling or jetting. Support to the borehole is provided by temporary casing or, for rotary drilling, a degradable polymer or other type of mud may be used. When the required depth is reached, the borehole should be cleared of drilling slurry. The well liner is then inserted into the hole and the filter media and any annular seals required are placed around the well as the casing is withdrawn. Certain filter materials may have to be placed by tremie (see Section 6.3.3).

A summary of the main techniques appropriate for installing dewatering wells is given in Table 2.9. Drilling techniques for water supply wells are discussed by Brandon (1986) and Sterrett (2009). Selection of well depth is considered in Section 6.3.2.

The bore diameter required for a well installation will depend on the outside diameter of the well liner and the annular thickness of any filter. In practice it is difficult to install a filter in an annulus less than about 50 mm wide. A filter thicker than 100 mm can lead to difficulties in developing the well (see Section 6.3.3). Centralisers on the screen are usually advisable to keep the thickness of the filter pack uniform. Whatever the drilling method, thorough flushing of the well with clean water (or clean mud) to remove drilling debris is essential before placing the screen and filter, especially in fine-grained soils. If a degradable polymer mud has been used, a chemical breaker may have to be poured into the well to encourage the breakdown of the mud. This may require the consent of the environmental regulatory authorities (see Section 4.2).

## Well development

See also
3.4.5 Clogging and biofouling
Chapter 4 Environmental matters
6.3.3 Filter design

In order to maximise the yield and to avoid damage to the submersible pump, wells should be developed before use. Where wells are in use for an extended period, yields can sometimes deteriorate as a result of clogging (see Section 3.4.5). Under these circumstances redevelopment of the well may be necessary periodically. The purpose of development is to:

- remove any residual drilling mud or debris from the filter pack or borehole wall, which might otherwise impair well efficiency
- increase the permeability of the aquifer in the immediate vicinity of the well by removing the finer soil particles (this is only viable in well-graded aquifers)
- yield clear water from the well, free of suspended solids
- remove any drilling or development debris from inside the well liner before installing the submersible pump.

*Table 2.9 Summary of principal drilling techniques used for dewatering well installation*

| Method | Resources | Typical diameter and depth of bore | Notes |
|---|---|---|---|
| Cable percussion (Figure 2.10) | Drill rig operator<br>Assistant drillers<br>Cable percussion drill and casing | 150 mm to 600 mm cased to about 50 m depth in unstable ground with casing telescoped<br>100 m depth or more in stable formations uncased | Widely available<br>Effective at penetrating granular and cohesive soils<br>Slow penetration if rock or cobbles and boulders present |
| Rotary open-hole with mud, direct circulation | Drill rig operator<br>Assistant driller<br>Rotary drill and rods<br>Mud<br>Mud handling system | 150 mm to 600 mm uncased to 100 m depth or more with appropriate rig | Rapid installation rates achievable in granular and cohesive soils<br>Cobbles and boulders can cause difficulty |
| Rotary open-hole with mud, reverse circulation | Drill rig operator<br>Assistant driller<br>Rotary drill and rods<br>Mud<br>Mud handling system | 400 mm plus uncased to 100 m depth or more with appropriate rig | Similar to direct circulation system, but usually used for larger holes |
| Rotary cased hole with water flush | Drill rig operator<br>Assistant driller<br>Jetting pump<br>Rotary drill and casing | 100 mm to 250 mm cased to 30 m depth or more with appropriate rig | An appropriate rig can penetrate virtually any ground from hard rock to soft clay |
| Rotary down the hole hammer | Drill rig operator<br>Assistant driller<br>Large compressor<br>Rotary drill and rods<br>Down the hole hammer<br>(Foam) | 76 mm to 600 mm to 100 m depth or more with appropriate rig | Requires the use of duplex systems in unstable formations<br>Can be quick and effective in hard rock |

Development involves alternately surging and pumping to achieve a flow reversal into and out of the well through the screen and filter pack. This washing action dislodges drilling debris and fine soil particles, flushing them into the well screen. For this procedure to be successful, the well screen aperture size and filter pack grading should be correctly sized and matched to the aquifer grading (see Section 6.3.3). In certain situations effective development can significantly improve the yield of a correctly specified and installed well, but no amount of development can recover the performance of an incorrectly specified or poorly installed well. Inappropriate development or the use of excessive energy during development can lead to a reduction in well performance or can even irrecoverably damage the well. For example, if the development process opens up a hole through a filter pack in a uniform fine-grained aquifer, continuous sand pumping could render the well useless. Some development methods are described in Box 2.3 and a detailed description of the development procedures used for water supply wells is given by Howsam *et al* (1995) and Sterrett (2009).

The discharge water arising from the well development process will contain suspended solids and possibly drilling mud. It may be feasible to remove some or most of this using a settlement tank. Discharge of water contaminated with suspended solids and drilling mud to surface waters can cause pollution, resulting in environmental damage and the possibility of prosecution by the regulatory authorities (see Chapter 4). As only a modest quantity of water arises from the development process, it is often possible to feed it on to the site surface or into a pit or sump to allow settlement of solids prior to discharge.

***Box 2.3 Summary of well development procedures***

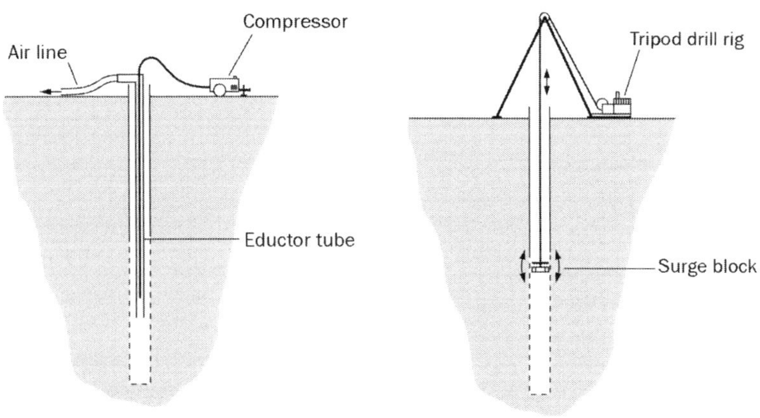


***a) Airlift with eductor tube*** ***b) Surge blocks***

**Air lift with eductor pipe:** using a compressor and an eductor pipe with a weighted air-line inside, the well can be pumped steadily to remove debris by air lift. Surging is achieved by lowering the air-line past the end of the eductor pipe and opening the air feed valve to blast the well. Air lift may not be feasible if the static groundwater level is too low; typically it should be no lower than about half the depth of the well.

**Surge block:** a block of slightly smaller diameter than the well liner is pulled sharply up a well using a tripod drill rig. As the block moves upward, a vacuum develops below the block drawing water into the well, and water is driven out of the well above the block. The debris that builds up in the base of the well needs to be removed periodically by bailing or air lift. The screen loadings developed with this technique can be very intense and it is not recommended for use in PVC liners unless thick wall screen is used.

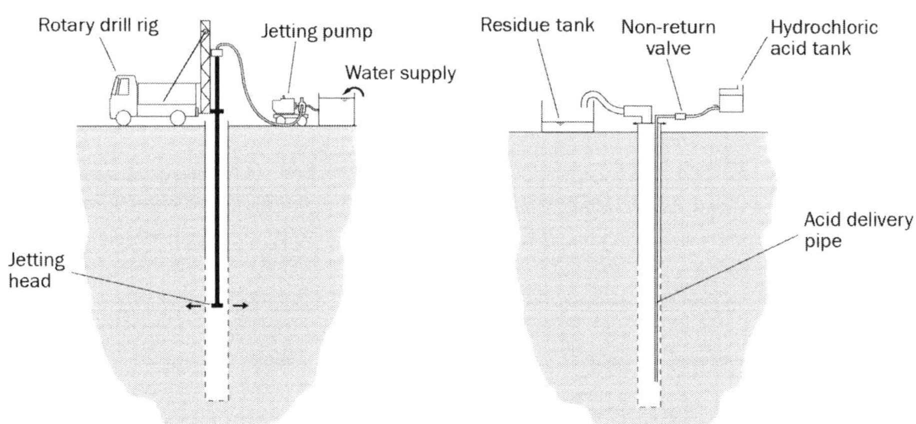


***a) Jetting*** ***b) Acidisation for chalk wells***

**Jetting:** a jetting head fitted with high-pressure horizontal water jets is passed over the screened section of the well. The jetting head is usually mounted on the end of the drill rods and is rotated as it is raised and lowered by a drill rig. The system may need to be alternated with air lift to achieve flow reversal and remove debris.

**Acidisation:** in carbonate rocks such as chalk, acid can be introduced into a well to dissolve any drilling slurry and possibly to open up the fissures in the aquifer around the well. Concentrated hydrochloric acid is used; the reaction releases large quantities of carbon dioxide, which may force acid from the well head unless appropriate precautions are taken. These works should be planned and carried out by experienced personnel so that appropriate health and safety measures are adopted.

## System layouts

See also
3.4.4 Standby facilities
6.5 Drawdown patterns

The basic principle for laying out a deepwell scheme is to space the wells evenly around the perimeter of the area where the control of groundwater is required (see Section 6.5). With deepwell systems the number of wells required for a scheme may be flexible. A few high-capacity wells or more smaller wells may give a similar extraction flow and drawdown. A few high-capacity wells may seem more cost-effective but, if there are uncertainties in the ground investigation information or a possibility of perched water, a larger number of smaller wells may give better control of the groundwater. Also, a scheme with too few wells may be unacceptable if the stoppage of a single pump could cause flooding or even catastrophic failure. Standby electric power supply facilities (Section 3.4.4) can be readily provided, but standby pumping plant ready for immediate start up is rarely provided for economic reasons. Typically, the solution is to make sure that there is sufficient redundancy in the pumping capacity and that the system is not highly dependent on any one well. This can be a problem for schemes that comprise less than three or four wells. There have to be sufficient wells to draw the water table down. Maintaining the lowered groundwater level may require a fewer number of wells and a reduced flow rate compared with the initial period of pumping.

## Deepwells used in conjunction with cofferdams

See also
2.1.9 Relief well

When wells are used to provide groundwater control for cofferdams, it has to be decided whether the wells should be located inside or outside the cofferdam. In practice deepwells often have to penetrate to a rather greater depth than the cofferdam retaining walls. This means that unless a natural geological cut-off (such as a clay layer) is present, flows and drawdown profiles may be very similar for internal and external wells. External wells have the advantage of being out of the way of the excavation and can reduce the hydrostatic loading on the retaining walls with potential for a saving in propping or anchoring requirements. Internal wells may benefit from some cut-off from the cofferdam. Internal wells also have a secondary potential benefit, which is that they can be set up to provide passive pressure relief (see Section 2.1.9) in the event of a total failure of the pumping plant or power supply system. This could be very important if the works involve pressure relief in an underlying confined aquifer, where failure to provide passive relief would lead to catastrophic failure of the excavation base.

# 2.1.6 Suction wells

See also
6.3.3 Filter design

A suction well consists of a deepwell, which is pumped by a surface suction pump, usually a wellpoint pump or a self-priming sump pump (Figure 2.17). Suction limitations of approximately 6 m are similar to those for a wellpoint system. As a result this arrangement is only likely to be suitable for drawdowns of 5 m to 6 m below the pump level. In appropriate circumstances this system can offer useful advantages:

- diesel pumps can be used so that no electrical power supply is necessary.
- diesel sump pumps are readily available and can be quickly mobilised and set up.
- installation of the well using cable percussion drilling techniques can penetrate ground, which is too permeable for wellpoint installation by jetting.
- because the well only has to accommodate the pump suction pipe, high yields are possible from relatively small diameter wells in appropriate soil conditions.

Suction wells are most appropriate for short-term shallow drawdowns in high permeability gravel aquifers. In these conditions wellpoint installation by jetting can prove difficult because of 'loss of boil' (see Section 2.1.3) and the capacity of deepwells may be limited by the size of readily available pumps. In a gravel aquifer it should be possible to use a coarse slotted well screen without a filter pack (see Section 6.3.3). In order to accommodate a 100 mm or 150 mm suction pipe, suction wells typically require a liner of at least 200 mm internal diameter.

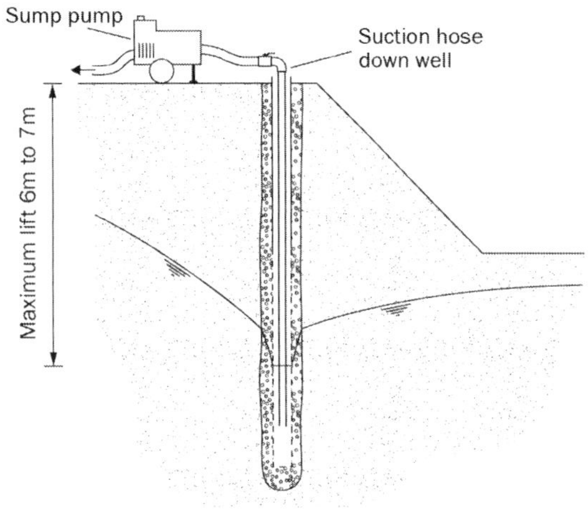


*Figure 2.17 A suction well*

## 2.1.7 Ejector wells

**See also**
2.2.3 Vacuum ejector wells
3.4.5 Clogging and biofouling

The ejector system, also known as the eductor system, consists of an array of wells pumped by jet pumps installed at the base of each well. Ejectors are generally used in one of two ways – in medium permeability soils in preference to a two-stage wellpoint system or a low flow rate deepwell system, or in low permeability soils to provide pore water pressure control by vacuum-assisted drainage. This section is concerned primarily with the former – the use of vacuum ejector wells for pore water pressure control is considered in Section 2.2.3.

Attributes of an ejector system are:

**Advantages:**

- operating depth is not limited by suction lift; ejectors are available with an operating depth down to 150 m, although most systems used for groundwater control purposes are limited to an operating depth of around 30 m to 50 m.
- ejectors will pump both air and water; this means that at low flows, if the well head and annulus is sealed, the ejector will develop a vacuum in the well, which can provide vacuum-assisted drainage in fine-grained soils.
- single-pipe ejectors can be installed in well liners as small as 50 mm internal diameter; this leads to a lower unit cost per well, allowing cost-effective installation of wells at close spacing if required.

**Limitations:**

- the capacity of individual ejectors is limited (see Box 2.4)
- ejectors have relatively low energy efficiency; this may not be a problem if total extraction flow rates are modest, but, for large flow rates, the power consumption can be prohibitive.
- ejector systems are sometimes susceptible to loss of performance from biofouling (Section 3.4.5) or nozzle and venturi wear; regular monitoring and maintenance is needed to identify any reduction in performance.

A typical ejector system layout identifying the main components is shown in Figure 2.18. The ejector body installed in the base of each well (Figure 2.19) contains a small diameter nozzle and venturi. The supply pipework feeds water from the supply pumps at high pressure, typically in excess of 700 kPa, to the nozzle. The supply flow passes through the nozzle at high velocity (up to 30 m/s), creating a pressure drop and generating a vacuum of up to 9.5 m of water at the ejector. This vacuum draws groundwater (the induced flow) through the well screen to the ejector body, where it joins the supply stream of water in the venturi and is piped back to ground level in the return riser pipe. The return water, which is the supply water plus the groundwater (the induced flow) is piped to a tank feeding the supply pumps and is recirculated back to the ejectors. The excess water abstracted from the ground builds up in the recirculation tank and is piped away to waste from an overflow. Ejector systems have two headermains, a supply headermain containing the high pressure feed to each well, and a return headermain to carry the recirculated water back to the supply pumps.

Two types of ejector bodies are available – twin-pipe ejectors and single-pipe ejectors. A schematic section of each type is shown in Figure 2.19. The twin-pipe ejector has separate supply and return risers and typically requires a well liner of approximately 100 mm diameter. The twin-pipe system has the advantage of performance flexibility, as a wider range of nozzle and venturi sizes can be accommodated. In the single-pipe system the supply and return pipe are arranged concentrically. The supply flow is fed down the annulus and the return feeds up the central pipe. The outer pipe can also be the well liner, providing it has sufficient pressure rating. This allows a single-pipe ejector body to be installed in a well liner of approximately 50 mm internal diameter.

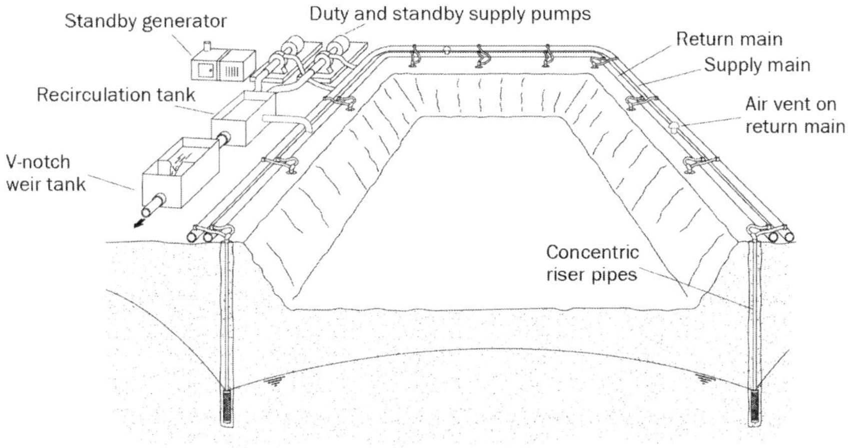


*Figure 2.18 Ejector system components*

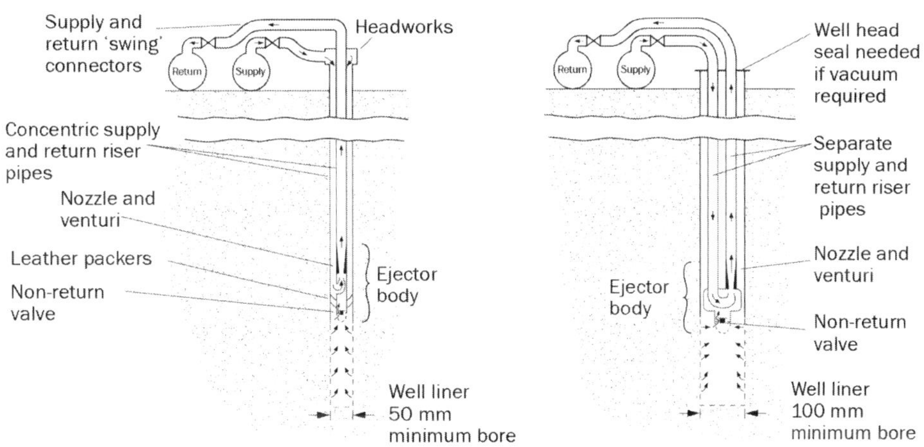


*a) Single pipe* *b) Twin pipe*

*Figure 2.19 Single-pipe and twin-pipe ejector bodies*

Ejector pipework is usually of PVC, HDPE, or steel and must be rated for the maximum operating pressures. Supply pumps are usually high speed single-stage or multi-stage rotary pumps. Supply pumps should be sized to drive the required number of ejectors in the system, taking account of the friction losses in the pipework.

## Performance of ejectors

The performance of an ejector is controlled by the following factors:

- design and geometry of the ejector
- size of the nozzle and venturi
- supply pressure at ground level
- depth of the ejector
- depth of water in the well above the ejector intake.

When designing a groundwater control scheme, it is normally necessary to determine the supply pressure and supply flow rate needed to obtain a given induced flow rate from an ejector operating at a certain depth in a borehole. In order to do this, performance curves of the form shown in Box 2.4 are required from the ejector manufacturer. Ejectors typically have the following characteristics:

1. The supply flow rate needed to obtain a given induced flow rate will increase with increases in the supply pressure at ground level and with greater ejector depths.
2. For a particular depth a minimum supply pressure is required to induce any flow; this is known as the stall pressure.
3. As the supply pressure increases beyond the stall pressure, the induced flow rate increases up to a maximum value when cavitation occurs. Any increase in supply pressure beyond that point will not increase induced flow rates.
4. Both the stall pressure and the supply pressure required to achieve cavitation increase with depth.
5. The maximum induced flow rate is independent of the ejector depth and the supply pressure, providing the supply pressure is sufficient to induce cavitation.

Ejectors are not damaged by the onset of cavitation and it is common practice to operate at or close to the cavitation point. This may be important where vacuum drainage is planned, because ejectors will only develop their maximum vacuum when cavitation occurs.

Further information on the performance of the ejectors in dewatering systems can be found in papers by Miller (1988), Powrie and Preene (1994b) and Siwec and White (1995).

# Ejector well specification and construction

**See also**
Table 2.7 Well liners

The range of well liners and screens available for ejector systems is essentially the same as for deepwells, as summarised in Table 2.7. For ejector wells the smaller sizes tend to be used, 50 mm to 104 mm internal diameter. Where single-pipe ejectors are to be used in a 50 mm well liner, the liner and liner joints must be rated to carry the intended supply pressure.

**See also**
Table 2.4 Installation methods
Table 2.9 Installation methods
6.3.3 Filter design

Wellpoint installation methods and deepwell drilling techniques summarised in Tables 2.4 and 2.9 are also applicable for ejector well installation. Ejectors are generally used in medium to low permeability soils, and for that reason careful attention has to be given to the screen and filter pack specification and installation to obtain optimum performance of the scheme (see Section 6.3.3).

*Box 2.4* *Performance curves for a single-pipe ejector*

For a typical single-pipe ejector, the relationship between depth, induced flow, supply flow and supply pressures is:

| Depth | Induced flow rate | Supply flow rate | Supply pressure |
|---|---|---|---|
| 10 m | 26 l/min | 28 l/min | 850 kPa |
| 20 m | 17 l/min | 29.5 l/min | 850 kPa |
| 30 m | 9 l/min | 31 l/min | 850 kPa |

**Ejector performance curves**

With most ejector designs it is possible to increase the induced flow rate by using a larger nozzle and venturi. Performance curves for two sizes of nozzle and venturi are shown here. Larger nozzle sizes will give greater induced flow rates at the expense of an increase in the supply flow (from Powrie and Preene, 1994b).

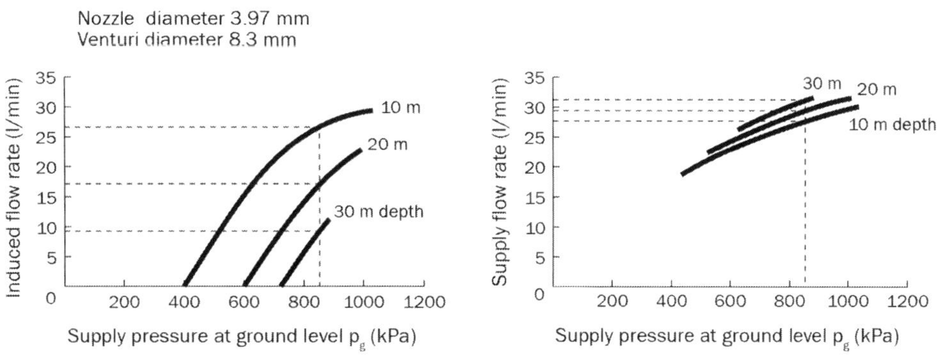


**Ejector performance curves for different nozzle sizes**

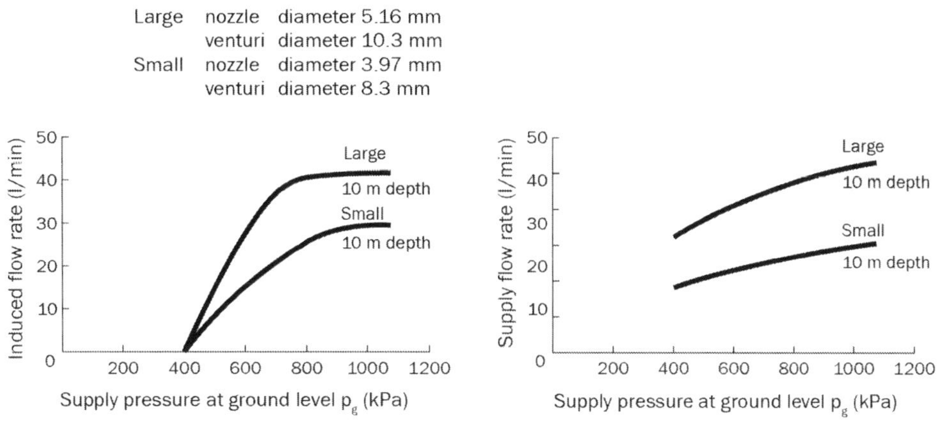


## System layouts

Like wellpoint and deepwell systems, ejector wells are generally laid out in a ring configuration around the area to be dewatered. Spacing of ejector wells will be controlled by the flow rate and the capacity of the ejectors used. If the soil stratification indicates the possibility of perched water or overbleed seepage, the well spacing may have to be reduced. In practice, ejector well spacings generally fall between those used for wellpoint systems, ie 1.5 m to 3 m, and those used for deepwells, ie 10 m or more. An example of the use of an ejector system for dewatering a shaft is given in McNamara *et al* (2008).

## Maintenance of ejector systems

See also
3.4.5 Clogging and biofouling

The important points to consider are:

1. Before it is switched on, the system should be primed with clean water and all pipework should be flushed out to avoid blockage of ejector nozzles.
2. Any suspended solids in the recirculating water can cause rapid wear of the nozzles. As the nozzles enlarge, the supply pressure will fall and the system performance deteriorate.
3. Biofouling (see Section 3.4.5) in the pipework can lead to a deterioration in ejector performance.
4. The yield from an individual ejector is determined by measuring both the supply flow rate and the return flow rate and taking the difference. For a given supply pressure and nozzle size, the supply flow should not vary. A regular record of supply flows (and return flows) can provide a useful indication of the onset of nozzle wear or biofouling.

# 2.1.8 Inclined wells

Inclined wells can be used to overcome limitations placed on well system layouts by surface access restrictions or underground services (examples are given in Box 2.5).

There are many situations where local departures from the optimum well spacing for a dewatering scheme will have relatively little impact on the system performance. If this is the case, it may be possible to use slightly deeper vertical wells or additional more remote wells to overcome access restrictions. However, there are situations where the restrictions are substantial or where even modest departures from the required well spacing may compromise the effectiveness of the groundwater control scheme. Where access is restricted, such problems are most likely to arise in the following situations:

- soils of low permeability where the cone of influence is steep and even minor seepage flow could cause ground loss
- stratified soils where perched water is present
- shallow aquifers where maximum drawdown is required to an underlying impermeable strata to minimise overbleed flow.

For inclinations up to about 10° from the vertical, very little modification to normal well installation procedure is required. For greater inclinations from the vertical, the method of placement of the well screen and any sand filter pack or grout seals, should be carefully considered. Technical advice from the pump manufacturer should be sought to ensure the pump can operate effectively at an angle.

# 2.1.9 Passive relief wells and sand drains

See also
Case study 7.4
Case study 7.12

It is sometimes possible to control excess pore water pressures in a confined aquifer below the base of a proposed excavation by using passive relief wells. The wells are drilled in the base of the excavation before the excavation has reached the piezometric level in the aquifer. As excavation continues below the piezometric surface, the wells will start to overflow, providing pressure relief. A schematic section of a passive relief scheme is shown in Figure 2.20.

Attributes of a passive relief well scheme are:

**Advantages:**

- the wells do not need to accommodate pumps and so can be of modest diameter; it may be possible to do away with the liner altogether and simply have a hole filled with sand or gravel
- water is removed using simple, robust and readily available sump pumping equipment rather than by deepwell pumps or ejectors.

**Limitations:**

- it can be difficult to prove the effectiveness of the system in advance of excavation unless some of the relief wells have liners installed and a pumping test is carried out.
- the passive relief wells feed water directly on to the excavation formation, which can lead to difficult working conditions if a network of collection drains is not maintained during excavation.
- relief wells can encourage softening of the strata immediately below the excavation
- relief wells can be difficult to seal on completion of the works.

In practice passive relief wells are generally only used for shafts or excavations in stable soils, stiff clay or weak rock, where there is only a marginal risk of base heave caused by sand lenses, fissures or a confined aquifer. The method is sometimes used as a permanent construction solution instead of providing floor anchorage (eg tension piles).

*Box 2.5 Case study of the application of inclined wells*

**Example 1**

Construction works for a new basement involved underpinning an adjacent building. Ground conditions consisted of water-bearing sandy gravels over stiff clay. In order to minimise overbleed at the interface between the gravel and clay, inclined wellpoints were installed at 1.5 m centres below the existing building, 'toed in' to the stiff clay (see figure). The residual seepage was dealt with by sump pumping.

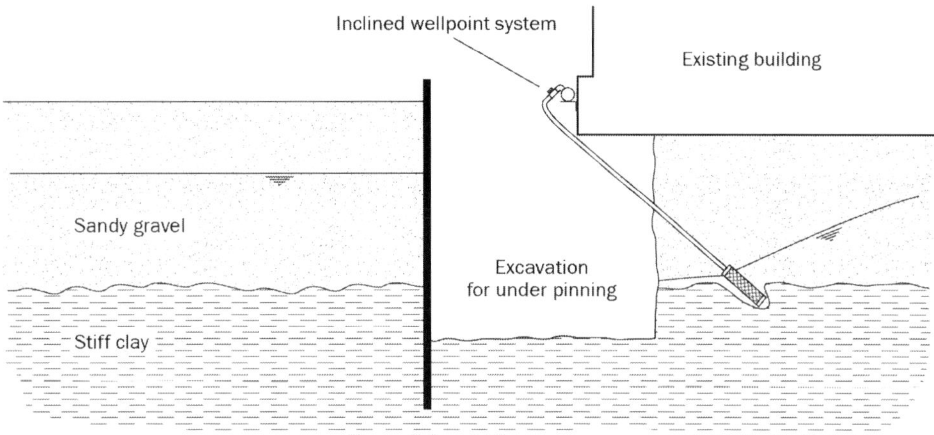


**Example 2**

Construction works for a railway underbridge involved jacking a pair of headings for the bridge footings beneath a railway embankment. Ground conditions consisted of water-bearing dense silty fine sand with possible silt and clay bands. A system of inclined ejector wells (see figure) was used to achieve a spacing of 2 m to 3 m below the railway embankment. This was necessary to achieve a satisfactory drawdown and to minimise the risk of overbleed above any silt or clay bands.

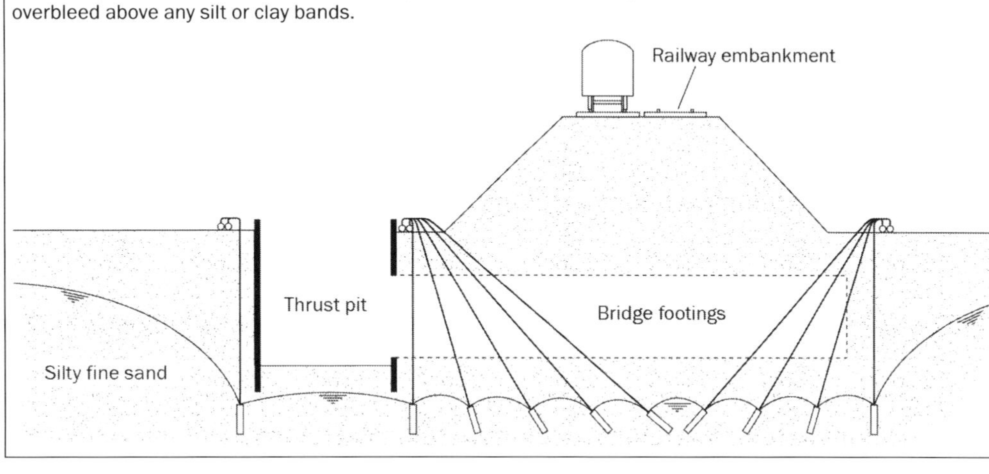

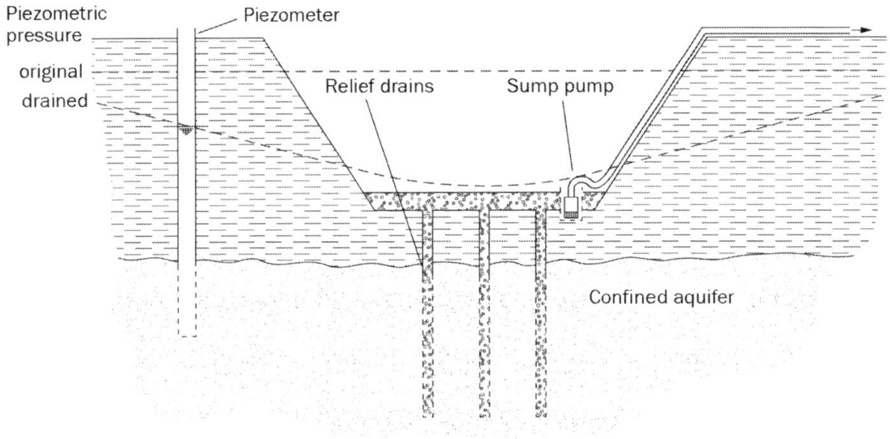


*Figure 2.20 Passive relief system*

Sand drains are a specific form of passive well that can be used to provide a hydraulic connection between two aquifers. As illustrated in Figure 2.21, this can be a useful method of draining a perched aquifer where an intervening clay layer may prevent groundwater from draining down to a lower aquifer, which is being dewatered. The water trapped in the upper aquifer can threaten the stability of the excavation unless it is drained (see Case study 7.4). Sand drains are holes formed by drilling, jetting or punching, which are then filled with sand or gravel of high permeability.

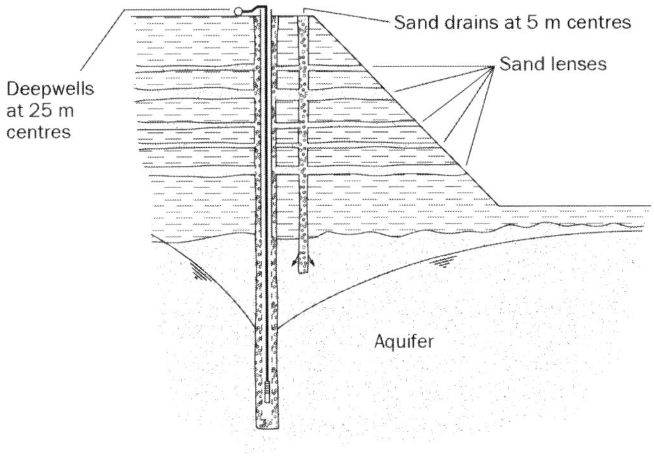


*Figure 2.21 Sand drain system*

Sand drains and other vertical drains (such as prefabricated band drains) are also used to reduce drainage paths to accelerate the consolidation of soft clays and silts beneath embankments (see ICE, 1982).

## 2.1.10 Tunnel and shaft dewatering

**See also**
Table 1.2 Cut-off methods
2.1.8 Inclined wells

For tunnels, shafts and adits between tunnels the same basic principles of groundwater control apply, but the access requirements and geometry are different from open excavations (see Powers *et al*, 2007, Chapter 23). A range of physical cut-off techniques can be employed for these works to control groundwater ingress (see Table 1.2). Dewatering can be avoided for tunnelling using certain types of full-face shields and for shafts constructed as flooded or 'wet' caissons with tremied concrete plugs cast underwater. However, situations do arise where control of groundwater by pumping from wells offers a cost-effective control technique.

Examples include:

- dewatering at shaft exit and entry 'eyes' for the launch or recovery of full-face tunnelling machines

- reduction in groundwater levels to reduce compressed air working pressures (health and safety risks and costs for compressed air working reduce appreciably at working pressures below 1 bar)
- for adit construction in water-bearing silts and fine-grained soils
- groundwater lowering to allow shaft sinking or open-face tunnelling in otherwise unstable ground
- groundwater control to aid recovery of a damaged or stuck tunnelling machine.

A spectacular example of the use of deepwells in conjunction with compressed air tunnelling is described by Biggart and Sternath (1996). In this case deepwells were installed into the sea bed to reduce pore water pressures to allow compressed air work to take place at less than 3 bar pressure. A smaller scale scheme using both surface wells installed in a river bed, tunnel wellpoints and compressed air is described in Case study 7.14.

Where the depth is not excessive and surface access is available, conventional installation of wells from the surface is often the most cost-effective technique, even if inclined well installation (see Section 2.1.8) is necessary (see Box 2.6). Alternatively, it may be possible to install wells directly out from a shaft or tunnel. A horizontal wellpoint scheme used for adit construction between a tunnel and shaft is described in Box 2.6. Installation of wells through a tunnel lining into unstable water-bearing ground is not a straightforward task. Some of the difficulties are:

- sealing the annulus between the tunnel lining and drill casing during well installation
- preventing soil from entering the drill casing during well installation
- controlling the drilling returns
- installing a sand filter pack
- sealing the annulus between the tunnel lining and well liner during pumping
- sealing the holes in the tunnel lining on completion.

In coarse well-graded soils persistent loss of ground should not occur through a narrow annulus and there may well be no requirement for a filter pack. Under these conditions steel wellpoints have been successfully installed in tunnel faces and through tunnel linings by a combination of jetting and jacking through cored holes.

In uniform fine sands and silts there can be substantial loss of ground in minutes from even a small hole or annulus of a few millimetres. Coring of the tunnel lining has to be carried out through a stuffing box securely bolted and sealed to the tunnel lining. Techniques for installing wells in these conditions include:

1. Drilling with a temporary casing and 'lost bit'. When the casing has reached full depth the bit is disengaged and the screen is installed as the casing is withdrawn.
2. Drilling with casing or polymer mud and fixed bit. As the drill string is withdrawn sand filter material in a polymer mud suspension is injected to keep the hole open. A well screen, usually of steel, is then pushed into the hole.

Successful well installation using these techniques requires careful planning, appropriate drilling equipment and experienced staff. In uniform fine-grained soils with excess groundwater heads of more than about 10 m, successful well installation can prove very difficult. An interesting description of the use of horizontal wellpoints in conjunction with both grouting and ground freezing for cross-passage construction between two tunnels is described by Doran *et al* (1995) and Biggart and Sternath (1996).

*Box 2.6 Case studies of tunnel and shaft dewatering*

**Example 1**

A short heading was required in glacial sands and gravels over bedrock. Surface access was restricted by an existing main road and services. Dewatering was carried out using vertical and inclined wells (see figure). Inclined wells were necessary because the bedrock limited the effective well depth so that wells were needed on both sides of the drive. Sump pumping was used to control the residual groundwater ingress at the face.

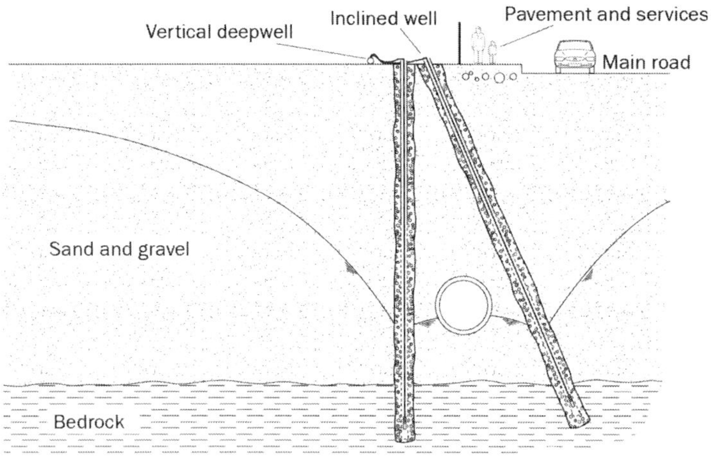


**Example 2**

An adit was to be constructed between a shaft and a tunnel. Ground conditions consisted of stiff clay but with a water-bearing fine sand layer 2 m thick at the level of the adit. The excess groundwater head in the sand layer was approximately 15 m. Horizontal wellpoints were installed from the tunnel and the shaft (see figure), which allowed adit construction by sprayed concrete lining (SCL) techniques. Wellpoints were needed on both sides of the tunnel to reduce the hydraulic pressure across the tunnel to avoid any risk of a blow.

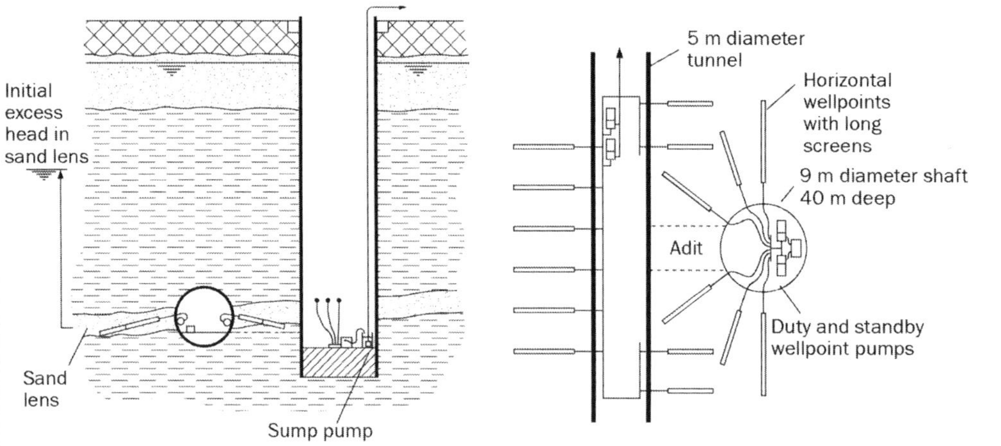


*a) Section* *b) Plan at tunnel level*

# 2.2 PORE WATER PRESSURE CONTROL SYSTEMS

## 2.2.1 Background

**See also**
Case study 7.3

In fine-grained soils of low permeability, such as silty sand or varved silts, the pore water pressures associated with even small quantities of water seeping into an excavation can cause serious instability. Side slopes may collapse or slump inwards and the base may become unstable or 'quicksand' conditions may develop. Conventional wellpoint or deepwell systems will yield little water and will probably not significantly improve the stability of the excavation. However, if a partial vacuum can be maintained in the wells, it is possible to achieve dramatic improvements in the stability of excavations (see Case study 7.3), even though well yields may not be substantially increased. This is because fine-grained soils cannot be literally dewatered, as their small pores will tend to remain saturated at negative pore water pressures. In a fine-grained soil the principal mechanism of drainage is consolidation rather than replacement of pore water by air. The aim of groundwater control in fine-grained soils is to reduce pore water pressures around an excavation, not to dry the soil out. The principal techniques used for pore water pressure control and the factors affecting their selection are summarised in Table 2.10. The techniques are described in detail in the following sections.

*Table 2.10 Pore water pressure control systems (after Preene and Powrie, 1994)*

| Technique | Advantages | Disadvantages |
|---|---|---|
| Vacuum wellpoints | Can pump relatively large flow rates | Drawdowns limited to 4 m to 6 m below headermain<br>Only limited vacuum can be developed in the well<br>Can be difficult to operate at very low flow rates |
| Ejector wells | Can develop vacuums of 9.5 m in the well<br>Drawdowns of 30 m to 50 m achievable | Flow capacity limited<br>Low energy efficiency<br>Can be prone to clogging by biofouling |
| Deepwells with vacuum | Can develop vacuums of up to 9.5 m in the well<br>Can pump relatively large flow rates<br>Drawdowns are theoretically unlimited | Two separate pumping systems are needed<br>Can be difficult to operate at very low flow rates |
| Electro-osmosis | Effective in very fine-grained soils<br>Can be used to enhance other techniques | Expensive because of high power consumption<br>Not commonly used so available expertise and experience limited |

## 2.2.2 Vacuum wellpoints

**See also**
2.1.3 Wellpoints

Relatively minor modifications to a conventional wellpoint system are required to make a vacuum wellpoint system. In a conventional system, described in Section 2.1.3, the vacuum is used to lift the groundwater up to ground level and into the pump intake. A conventional wellpoint system can achieve a maximum lift of about 6 m below the headermain level. In a vacuum system some of the vacuum is used to lift the water and some to maintain the wellpoint filter column at below atmospheric pressure. This is achieved by limiting the suction lift to less than 6 m and by sealing the wellpoint filter column (Figure 2.22a).

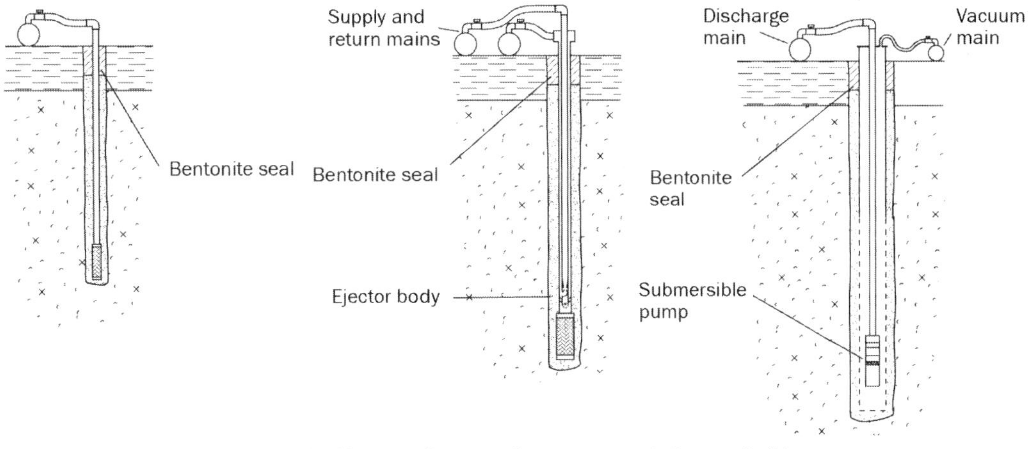


*Figure 2.22 Vacuum-assisted dewatering systems*

The wellpoint is installed by jetting or drilling and surrounded by filter material (see Section 2.1.3) and the top of the borehole is sealed with a clay or bentonite plug. The plug prevents air entering the filter medium, allowing a vacuum to be developed in the whole filter column. Even if extra vacuum pumps are used and great care is taken to avoid air leaks in the vacuum system, drawdowns are normally limited to between 3 m and 4.5 m below header pipe level. If greater drawdowns are required, multi-stage vacuum wellpoint systems can be used (see Figure 2.5), but in such cases one of the other pore water pressure control techniques should be considered.

The design and operating procedure for a vacuum system are essentially the same as for conventional wellpoint systems. Wellpoint spacing for vacuum systems are generally in the range 1.5 m to 2 m for soils of low permeability (see Table 2.3).

## 2.2.3 Vacuum ejector wells

**See also**
2.1.7 Ejector wells

Ejectors are ideally suited to pore water pressure control in fine-grained soils. An ejector will pump both air and water, so if the well filter and well casing are sealed, a vacuum will automatically be developed. The low well yields from fine-grained soils are suitable for the flow characteristics of ejectors, which cannot cope with high flow rates. A typical ejector well installation for pore water pressure control is shown in Figure 2.22b. With a single-pipe ejector installed in a 50 mm diameter well liner, the casing is effectively sealed and it is only necessary to add a clay or bentonite plug to the filter column. Design and installation procedures for ejector systems are described in Section 2.1.7.

Ejectors are capable of generating a vacuum of about 9.5 m of water in a sealed well. Ejectors are available with operating depths down to 150 m, although in practice most systems used for dewatering works are limited to about 30 m to 50 m. Ejector wells can be economically installed at a spacing of 3 m – for most pore water pressure control applications a spacing in the range 3 m to 15 m is used. Examples of the use of an ejector system for pore water pressure control are given by Powrie and Roberts (1990) and Roberts *et al* (2007).

## 2.2.4 Deepwells with vacuum

**See also**
2.1.5 Deepwells

A conventional deepwell system can be enhanced to provide pore water pressure control by sealing the well casing and filter column, and evacuating the well using a vacuum pump at ground level. This arrangement is illustrated in Figure 2.22c. Design and installation procedures are the same as for conventional deepwell schemes (see Section 2.1.5). Vacuum is provided by an exhauster unit, which is usually electrically powered. The vacuum pipework can be of relatively small diameter, say 25 mm to 76 mm bore, because once the vacuum is established, air flows should be low. The seal on the well casing has to accommodate the pump riser pipe, power cable and the vacuum connection. A vacuum gauge is also useful.

Submersible pumps for deepwell systems with vacuum need an allowance of an additional 10 m on the discharge head to overcome the vacuum in the casing. Slim-line borehole pumps rely on a flow of water to cool the electric motor and lubricate the bearings, so difficulties can arise when deepwell pumps are run at very low flow rates. These can be overcome by the use of well water level controllers, which stop and re-start the pump, although it complicates the electrical control system, especially on a large scheme.

## 2.2.5 Electro-osmosis

Electro-osmosis involves setting up a direct electric current between electrodes placed in the ground to induce flow of the positively charged ions surrounding the soil particles, along with the pore water, from the anode to the cathode. The water is collected at the cathode and pumped away, usually by wellpoints or ejectors. A schematic diagram of the process is shown in Figure 2.23.

The principles of electro-osmosis were developed by Leo Casagrande in the 1930s and the technique has only occasionally been applied around the world since then. The development of the process and some early applications are described by Casagrande (1952). Case studies of more recent applications are given by Casagrande *et al* (1981) and Doran *et al* (1995).

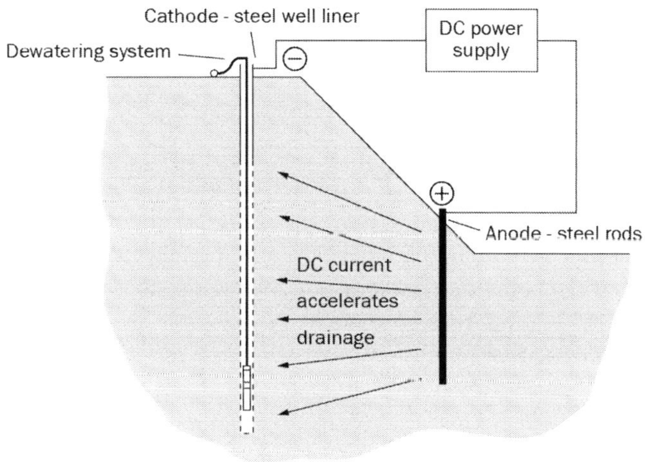


*Figure 2.23 Principles of electro-osmosis*

Electro-osmosis can be used to provide effective pore water pressure control in very fine soft silt and clay soils, which are at or beyond the lower permeability limit for vacuum-assisted drainage. The application of the technique is constrained by the high cost of the heavy power consumption and by the health and safety aspects of using direct electric currents in the ground on a construction site.

# 2.3 GROUNDWATER RECHARGE SYSTEMS

## 2.3.1 Background

**See also**
4.3 Regulatory framework
Case study 7.9
Case study 7.11

The concept of artificial recharge is that water is returned to the ground around the site to prevent groundwater levels falling outside prescribed limits. The recharge water is usually the water abstracted by the groundwater control system, although mains water is sometimes used. In addition to controlling groundwater levels, recharge systems are sometimes considered as a means of disposing of the groundwater abstracted by the dewatering system (see Case studies 7.9 and 7.11). Caution should be exercised when considering a recharge scheme to control groundwater levels or as a means of discharge disposal; such schemes are complex to operate and monitor and require careful planning. The artificial recharge system can become a recharge boundary and result in the recirculation of water leading to reduced efficiency of the groundwater control system. As such, the proximity of the artificial recharge needs to be taken into account when designing the groundwater control system. However, subject to obtaining the necessary regulatory permissions, artificial recharge systems are a potential mitigation measure to control the impact on sensitive water-dependent features, such as wetlands, lakes, watercourses and water supply boreholes.

A groundwater control scheme may generate drawdowns around a site, which are unacceptable (see Powers, 1985, and Preene and Brassington, 2003). For example:

- where drawdown could lead to drainage of a loose or soft stratum that would result in unacceptable consolidation and surface settlements
- where a water supply well is present within the distance of influence of the groundwater control scheme and the drawdown could cause derogation of the supply
- where drawdown could lead to leaching or spreading of contaminants already present in the vicinity of the groundwater control scheme
- where drawdown could lead to saline intrusion into a coastal aquifer
- where drawdown could cause old timber piles to dry out, exposing them to the risk of rapid deterioration from aerobic attack. This risk arises in Copenhagen, Denmark and is described by Raben-Levetzau *et al* (2004) and by Bock and Markussen (2007).

Off-site drawdowns can be controlled either by physical cut-offs around a site or by an artificial recharge scheme.

Recharge of groundwater is generally more difficult than abstraction. Recharge wells are prone to clogging by even small quantities of suspended solids or precipitates in the recharge water. Air entrainment in the discharge water can also lead to clogging as the trapped air becomes a physical barrier to flow within the recharge well. As a rule of thumb, for each abstraction well two or three recharge wells may be required when abstracting and recharging into the same stratum. This is to allow for sufficient capacity and for a number of the wells to be out of commission being rehabilitated. In practice many recharge schemes use a combination of internal dewatering, a partial cut-off and an external recharge system to reduce flows and to avoid the potentially extensive well arrays needed to recharge into the same stratum. An interesting case study of the use of a slurry wall and recharge via sand drains to control settlement in a low permeability setting is described in Ervin and Morgan (2001).

**See also**
6.6 Settlement

The feasibility of an artificial recharge scheme and the cost of any alternative solutions should be examined carefully. Some settlement may be acceptable, and in any case untoward settlements are unlikely if drawdowns are kept within seasonal fluctuations of the groundwater levels or do not exceed historic drawdowns where pumping has taken place beneath urban areas. Methods of assessing settlements are considered in Section 6.6.

Good quality site investigation information, ideally including pumping test and groundwater chemistry data, is essential to assess the viability of artificial recharge. The design basis for a recharge system, which included an abstraction recharge trial and numerical modelling, is described by Roberts and Holmes (2011) (see Case study 7.11). As well as clogging, recirculation may be a problem. If recharge is attempted too close to a dewatering system, the extraction flow rate may have to be increased in order to maintain the drawdown, leading to an increase in the scale of the recharge scheme; a vicious circle may result. In order to minimise the effects of recirculation, recharge is often carried out at one to two times the distance of influence of the dewatering system from the site. For large drawdowns in medium to high permeability aquifers, the distance of influence may be several hundred metres or even a few kilometres. In certain situations this problem can be overcome by the use of partial cut-offs. An example of a recharge system, which successfully exploited a partial cut-off, is given in Box 2.7.

*Box 2.7 Case study of a recharge system with partial cut-off (after Powrie and Roberts, 1995)*

A system of recharge wells was installed to minimise external drawdowns during dewatering works for a deep basement at a city centre site. External drawdowns could have caused underdrainage and consolidation of a superficial layer of alluvial loam and peat 5 m thick. A number of listed historic buildings (including a cathedral) with a history of settlement damage were present near the site. Seepage flows from the highly permeable gravel stratum were excluded by a deep diaphragm wall. Flows from the underlying chalk were controlled by a system of 20 internal deepwells screened in the chalk, pumping approximately 100 l/s in total. External drawdowns were kept within acceptable limits by recharging 50 to 60 per cent of this flow via 10 external deepwells screened in the gravel (see figure). Recharge into the gravel required relatively few recharge wells because the gravels were significantly more permeable than the chalk.

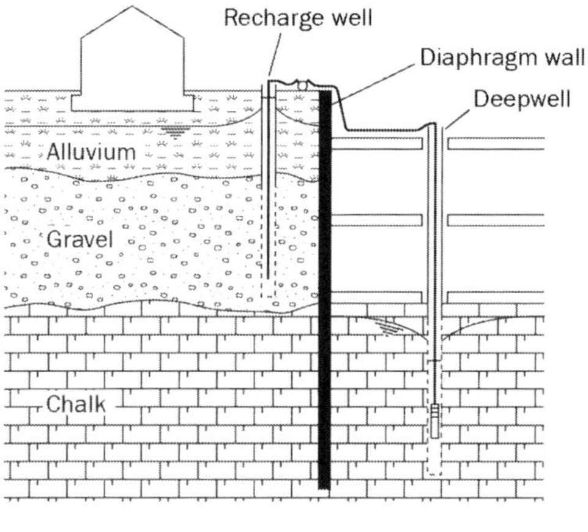


The operation of any form of recharge system requires a discharge consent from the environmental regulatory authorities (see Section 4.2). This provision applies even if the groundwater is being abstracted and returned to the same aquifer on the same site.

## 2.3.2 Recharge trenches

Recharge trenches have to be excavated to penetrate through any superficial low permeability deposits. The trenches are kept topped up with water and infiltration occurs out of the base of the trench. A section through a trench recharge system is shown in Figure 2.24.

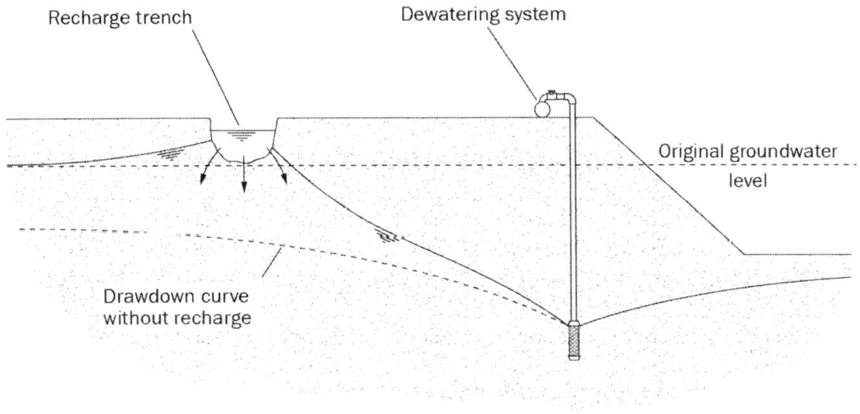


*Figure 2.24 Trench recharge system*

Satisfactory control of groundwater levels using recharge trenches is difficult for a number of reasons:

- Flows from recharge trenches cannot quickly be adjusted or turned off.
- The amount of water flowing out of a recharge trench cannot be determined quickly.
- The base of recharge trenches often become clogged and may require periodic cleaning out with an excavator.
- If the base of the trench is significantly above the standing groundwater level, the effect of the infiltration on groundwater levels can be very unpredictable.

Recharge trenches are used to good effect for irrigation and sometimes in the water supply industry. However, for construction dewatering schemes the combination of poor control and poor predictability, plus the extensive space potentially required, severely constrains their use.

## 2.3.3 Recharge wells

**See also**
2.1.5 Deepwells
3.4.5 Clogging and biofouling

Unlike recharge trenches, recharge wells can be designed to inject water at a specific level in the sequence of stratification, and the feed pipework can be set up to give good flow control and allow accurate performance monitoring. The hydraulic requirements for recharge wells are essentially the same as for extraction wells. Both need to be as efficient as possible, with minimum well losses. As a result, recharge wells are designed, drilled and developed in exactly the same way as extraction wells (see Section 2.1.5). The only difference is that recharge wells do not need to accommodate a pump, so, for the same flow rate, recharge well liners may be of smaller diameter. In operation extraction wells are self-cleaning and redevelopment is only necessary when there is evidence of biofouling or clogging (see Section 3.4.5). Recharge wells, on the other hand, are very prone to clogging and, unless the recharge water is of excellent quality, regular redevelopment may be necessary.

A typical recharge well set up is shown in Figure 2.25. Air vents are required at high points in the feed pipework to avoid air locks. A down spout is essential to prevent the recharge water from cascading into the well. Cascading can promote biofouling and can cause entrained air to be forced into the aquifer, restricting recharge flows. It is good practice for the feed pipework to include a meter to monitor recharge flows. Recharge flow rates combined with measurements of the water level in the well allow the performance of the well to be monitored so that the need for redevelopment can be assessed.

If the standing groundwater level is relatively high, the recharge wells will almost certainly require a substantial grout seal to prevent water short-circuiting up the filter pack to ground level. If necessary, the well head can be sealed and the recharge pipework pressurised slightly so that the feed head is 2 m or 3 m above ground level. In order to avoid over-pressurising the well, a constant head tank or a standpipe, which can overflow, should be used to provide the feed head.

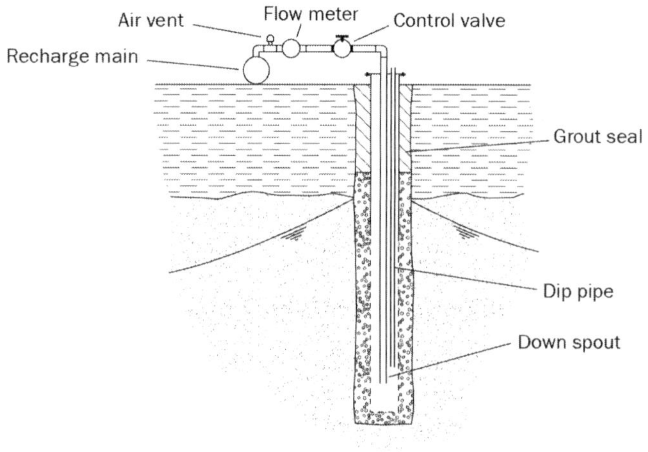


*Figure 2.25 Recharge well*

The importance of the feed water quality to the success of a recharge operation cannot be overemphasised. There are a number of problems with feed water quality:

1 Fine or colloidal particles can lead to rapid clogging of wells.

2 Abstracted groundwater may contain dissolved iron. In aerobic conditions, insoluble iron-based compounds will precipitate and biofouling may occur (see Section 3.4.5). Box 2.8 shows that the resulting clogging can be severe.

3 Both abstracted groundwater and mains water may contain dissolved air or methane, which can be released as the pressure falls (or the temperature rises) in the feed pipework. The bubbles can then be driven into the formation and cause clogging. Degassing equipment has been used to overcome this problem (Rijkswaterstaat, 1986).

4 Recharge water from the mains or from a different aquifer may be incompatible with the groundwater, resulting in chemical precipitation and clogging.

Where clogging does occur, mitigation measures should be adopted. A programme of regular well development or cleaning may be needed, as described in Box 2.8. Well redevelopment will not always recover the full capacity of a recharge well and, in some circumstances where wells are in operation for very long periods, recharge wells may need to be replaced as the overall system capacity falls.

If recharge is required into an aquifer of medium to low permeability, a recharge wellpoint system could be considered. Design and installation are the same as for a conventional wellpoint system (see Section 2.1.3).

***Box 2.8 Case study of recharge system with iron-related biofouling***

A groundwater control system consisted of 10 abstraction wells. Recharge was required to prevent depletion of the underlying chalk aquifer (see Box 4.4), so 30 recharge wells were installed between 500 m and 1000 m from the abstraction system. The groundwater contained 2 to 5 mg/l of dissolved iron; in operation the recharge wells clogged up within a few days because of biofouling by Gallionella bacteria, which reduced the capacity of an individual well by more than 75 per cent. The system was able to function satisfactorily because sufficient recharge wells were provided to allow a number of them to be out of commission for regular cleaning. The total recharge flow could be handled by between 20 and 25 unclogged wells, so at any one time 5 to 10 of the wells could be disconnected from the system to allow the biofouling to be removed by flushing with compressed air. As each well was cleaned, it was reconnected to the system and another well was disconnected. In this way a cleaning cycle was set up so that every recharge well was cleaned approximately once a week.

## 2.4 KEY REFERENCES

CASHMAN, P M and PREENE, M (2012) *Groundwater lowering in construction: a practical guide to dewatering, second edition*, CRC Press, Boca Raton, USA (ISBN: 978-0-41921-110-5)

POWERS, J P (1985) *Dewatering – avoiding its unwanted side effects*, American Society of Civil Engineers, New York, USA (ISBN: 978-0-87262-459-7)

POWERS, J P, CORWIN, A B, SCHMALL, P C and KAECK, W E (2007) *Construction dewatering and groundwater control: new methods and applications, third edition*, Wiley-Blackwell, New York, USA (ISBN: 978-0-47147-943-7)

PREENE, M and POWRIE, W (1994) "Construction dewatering in low permeability soils: some problems and solutions" *Proceedings of the ICE – Geotechnical Engineering*, vol 107, 1, Institution of Civil Engineers, London, pp 17–26

RIJKSWATERSTAAT (1986) *Groundwater infiltration with bored wells*, Rijkswaterstaat Communications 39, The Hague, The Netherlands

# 3 Operation and management

This chapter provides guidance on matters relating to health and safety, forms of contract, site operations and monitoring. The guidance on health and safety is restricted to matters relating to groundwater control. Broader-based advice on health and safety in the construction industry can be found in publications by CIRIA and by the Health and Safety Executive (HSE, 2006).

## 3.1 HEALTH AND SAFETY REGULATIONS

The main regulations covering occupational health and safety in the building and construction industries include:

- Health and Safety at Work Act 1974.
- Management of Health and Safety at Work Regulations 1992.
- Construction (Design and Management) Regulations 2015.

The regulations listed here and elsewhere in this document, are those in force as of October 2015. Regulations may be updated from time to time, and the reader should check to ensure they comply with the latest regulations.

These regulations control the practical ways in which construction and building work is carried out on site. The regulations particularly relevant to groundwater control operations on site are listed in Table 3.1. In addition, reference should be made to guidance by the British Drilling Association (BDA) on general safety policy, risk analysis and method statements for drilling on sites (BDA, 2008 and 2015).

*Table 3.1 Health and safety regulations particularly relevant to groundwater control operations on site*

| Legislation | Main provisions |
|---|---|
| Provision and Use of Work Equipment Regulations (PUWER) 1998 | Machinery protective guards and controls, including requirements for guarding of drilling rigs |
| Electricity at Work Regulations 1989 | Maintenance of equipment, certification and training of operatives |
| Work at Height Regulations 2005 | Avoidance/reduction of work at height, use of work equipment to reduce the risk and consequence of falls |
| Construction (Head Protection) Regulations 1989<br>Personal Protective Equipment at Work Regulations 1992 | Head protection for each employee, maintained and replaced as necessary. To be worn unless there is no foreseeable risk of head injury |
| The Control of Noise at Work Regulations 2005 | Reduction of noise levels and the use of hearing protection |
| Control of Substances Hazardous to Health Regulations (COSHH) 2004 | Assessment and control of all hazardous substances, records of hazardous substances held by a principal contractor, instructions in the hazards and precautions to be followed |
| The Regulatory Reform (Fire Safety) Order 2005 England and Wales. Fire (Scotland) Act 2005 | Fire safety |
| Lifting Operations and Lifting Equipment Regulations (LOLER) 1998 | Certification of lifting equipment, training of competent operatives |

# 3.2 CDM REGULATIONS 2015

## 3.2.1 Background and regulations

The Construction (Design and Management) Regulations 2015 (CDM 2015) are part of the health and safety legislation that places duties upon clients, designers and contractors to think through their planning and management of health and safety. The CDM Regulations and the associated approved code of practice were introduced in the mid-1990s to extend the traditional health and safety responsibilities of contractors to include clients and designers; they were subsequently updated in 2007 and 2015. Apart from certain exemptions set out in CDM 2015, there is a requirement that health and safety be considered throughout the life of a project from design to construction, maintenance and demolition.

A practical guide to the responsibilities and application of CDM 2015 is provided in HSE (2015). A brief description of the duties and requirements is as follows:

### Clients

Responsible to make suitable arrangements for managing the health and safety of a project including the appointment of other duty holders, ensuring sufficient time and resources are allocated, ensuring relevant information is prepared and provided to other duty holders, and ensuring that the principal designer and principal contractor carry out their duties.

### Principal designers

Responsible to plan, manage, monitor and coordinate health and safety in the pre-construction phase of a project including management of foreseeable risks, ensuring designers carry out their duties; preparing and providing relevant information to other duty holders, and liaising with the principal contractor to help with the planning management, monitoring and coordination of the construction phase.

### Principal contractors

Responsible to plan manage, monitor and coordinate the construction phase of a project including liaising with the client and principal designer, preparing the construction phase plan; and organising cooperation between contractors and coordinating their work.

Principal contractors must ensure that suitable site inductions are provided, reasonable steps are taken to prevent unauthorised access, workers are consulted and engaged in securing their health and safety, and that welfare facilities are provided.

### Construction phase plan

During the pre-construction phase, and before setting up a construction site, the principal contractor must draw up a construction phase plan or make arrangements for one to be drawn up. The plan must set out the health and safety arrangements and site rules taking into account industrial activities taking place on the site. The principal designer must assist the principal contractor in preparing the construction phase plan by providing to the principal contractor all information the principal designer holds that is relevant to the construction phase plan. Throughout the project the principal contractor must ensure that the construction phase plan is appropriately reviewed, updated and revised from time to time so that it continues to be sufficient to ensure that construction work is carried out, so far as is reasonably practicable, without risks to health and safety.

### Health and safety file

During the pre-construction phase the principal designer must prepare a health and safety file appropriate to the characteristics of the project which must contain information relating to the project that is likely to be needed during any subsequent project to ensure the health and safety of any person. The principal designer must ensure that the health and safety file is appropriately reviewed, updated and revised from time to time to take account of the work and any changes that have occurred. During the project the principal contractor must provide the principal designer with any information in the principal contractor's possession relevant to the health and safety file, for inclusion in the health and safety file. At the end of the project the principal designer, or, where there is no principal designer, the principal contractor must pass the health and safety file to the client.

## 3.2.2 Application of CDM 2015 to groundwater control

The way in which CDM 2015 are usually applied is by considering the different stages of a project from feasibility to design and construction. Risk assessments are carried out for each stage. Further guidance is given in two publications, CIRIA C755 (Ove Arup & Partners and Gilbertson, 2015) and CIRIA C756 (Gilbertson, 2015).

The aim is to prevent potential hazards or protect against them by:

- avoiding foreseeable risks
- reducing risks at source
- giving priority to measures, which will protect all persons affected by the works rather than just the individual at work.

### Site investigation

**See also**
Chapter 5 Site investigation

Without a site investigation, which adequately addresses the information needs of all parties, such as designers and contractors, assessment and control of many potential hazards will be difficult. Chapter 5 discusses the specific points, which must be considered when designing and procuring a site investigation for a project where groundwater control may be required.

### Feasibility study

Strategic decisions taken at the feasibility stage can have a major impact on health and safety on site. The most fundamental decision to be taken is whether an excavation, eg for a basement, tunnel, or shaft, is necessary. Once that is decided, groundwater control is an integral part of the design process. The permanent works designer should consider the impact possible groundwater control measures may have on the design and make any necessary allowances or alterations.

### Design and planning phase

**See also**
Chapter 5 Site investigation

As part of the temporary works, the design of groundwater control measures is often undertaken by the principal contractor or a specialist contractor. Under CDM 2015 the temporary works designer has the same obligations as the permanent works designer. The designer has to consider potential hazards associated with groundwater control, which could have been reasonably foreseen, avoided or reduced. The risk assessment will involve gathering further information about the site from a desk study and ground investigations (preliminary and main), as explained in Chapter 5. A record of the site investigation is kept in the health and safety file and used in the preparation of the construction phase plan.

Some examples of potential hazards with possible preventative or protective measures are given in Ove Arup & Partners and Gilbertson (2015) and Gilbertson (2015) Examples relevant to groundwater control are given in Table 3.2.

*Table 3.2 Examples of potential hazards and preventative or protective measures*

| Potential hazards | Preventative or protective measures |
|---|---|
| *Infrastructure* | |
| Risk from buried services | Locate services from documents<br>Excavate hand-dug starter pits to check for or expose services |
| Structural damage to buried services or adjacent buildings caused by excessive ground movement | Relocate or reroute the works or services<br>Limit extent of drawdown and maximum allowable settlement<br>Employ pumping methods in combination with exclusion techniques (see Section 1.2.7) |
| *Geotechnical conditions* | |
| Variable stratigraphy of low and high permeability strata | Design the groundwater control measures to control pore water pressures for discrete zones, possibly using a combination of methods |
| Flooding of excavation from surface | Provide adequate surface water and seepage control (eg drains and sump pumps) |
| Heave of excavation floor or quicksand conditions | Control pore water pressures at depth to limit upward hydraulic gradients |
| Collapse or slumping of excavation slopes and faces | Control pore water pressures in the area of slopes or retaining walls<br>Avoid excessive hydraulic gradients to reduce the risk of localised erosion |
| Flooding of excavation due to failure of duty pumping system | Provide adequate standby pumping plant and power supply |
| *Past use* | |
| Contaminated soil and groundwater | Relocate or reroute the works<br>Install cut-off barrier to control migration of contaminants<br>Provide for on-site treatment of discharge water |

In addition to potential hazards, there are likely to be practical or financial constraints that affect the choice of groundwater control methods. These include:

- depth and area of the excavation, eg will the size and geometry of the excavation affect the need for support and the suitability of the groundwater control methods?
- access to the site, eg are there space restrictions that could limit the choice of method or plant?
- programme requirements, eg could the programme affect the choice of methods?
- cost constraints, eg is cost of primary or secondary importance?
- effectiveness of the method, eg are minor or localised seepages or inflows acceptable?

### Construction phase

During construction the responsibility for the construction phase plan is with the principal contractor. Appropriate method statements should be produced for the specific groundwater control measures to be used. The groundwater control operations may have to be modified for a variety of reasons (such as unforeseen ground conditions). The changes will be recorded in the health and safety file and used to modify the construction phase plan.

## 3.2.3 Construction phase plan

There is unlikely to be a construction phase plan specifically for groundwater control operations. The identification and assessment of potential hazards will most likely be carried out for the project as a whole. Ove Arup & Partners and Gilbertson (2015) and Gilbertson (2015) includes examples that illustrate construction phase plans for a range of construction projects.

In order to prepare and develop the construction phase plan, the principal contractor will use the pre-construction information. The principal designer has the responsibility to plan, manage, monitor and coordinate health and safety in the pre-construction phase of a project including management of foreseeable risks, ensuring designers carry out their duties, preparing and providing relevant information to other duty holders, and liaising with the principal contractor to help with in the planning management, monitoring and coordination of the construction phase. The type of information would include:

- **Site investigations.** Natural and man-made ground conditions that could pose a risk to health and safety during the construction phase should be identified, eg buried services, water abstraction boreholes, contaminated land and adjacent properties.
- **Principles of design.** Although the design of a groundwater control system is likely to be undertaken by the principal contractor or specialist contractor, the designer is still obliged to make clear the principles of the design and describe any special requirements for the purpose of construction. These may include a geotechnical assessment of slope stability for an excavation and the suitability of dewatering techniques compared with exclusion techniques.

# 3.3 CONTRACTUAL MATTERS

## 3.3.1 Background

See also
3.4 Monitoring
Chapter 4 Environmental matters

Groundwater has the dubious distinction of being a frequent cause of disputes in construction projects. Even if a comprehensive site investigation is carried out, there will remain a risk that a dewatering system will not provide adequate control of groundwater. Dewatering works are often needed during the early stages of construction on a project and many subsequent activities may depend on the effective control of groundwater. Consequently, the control of groundwater for temporary works is sometimes seen as a high risk operation with the potential for significant cost overruns. This image is partly confirmed by Roberts and Deed (1994), who examined records from over 130 groundwater control contracts and found average cost overruns of 35 per cent, with a doubling of costs not uncommon (see Figure 3.1).

The main cause of cost overrun was identified as the extension of the period of pumping (as a result of project delays unrelated to groundwater control) rather than unforeseen ground conditions. In fact, unforeseen ground conditions were only found to be a factor in 8 per cent of the projects examined.

Where unforeseen ground conditions are a factor, the costs resulting from delay and disruption can be substantially greater than the direct increase in the cost of the groundwater control works. Monitoring during installation and initial drawdown (see Section 3.4) can allow a prompt response to unforeseen conditions, thereby minimising any delay and disruptions to the works.

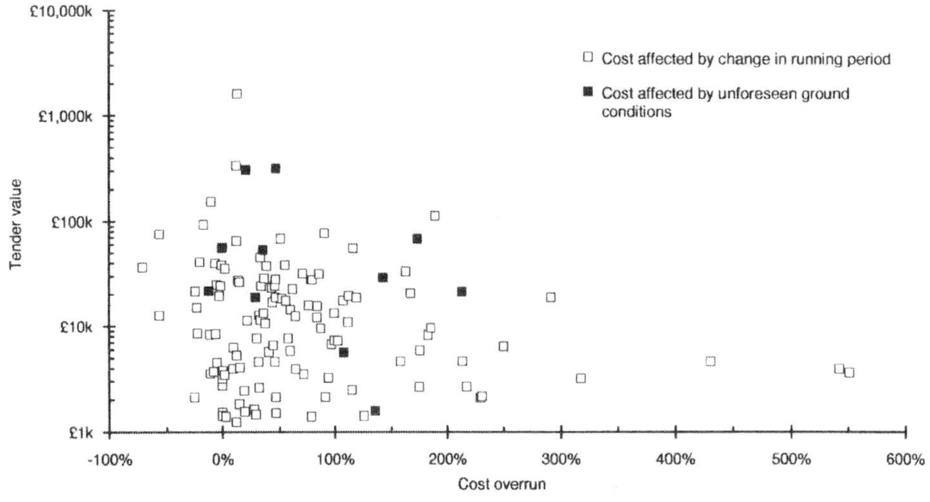


*Figure 3.1 Tender value versus cost overrun for dewatering subcontracts (after Roberts and Deed, 1994)*

Although the percentage cost overruns can appear high, the cost of groundwater control operations may be small in relation to overall project costs. The tender cost for groundwater control rarely exceeds about 1 per cent of total costs for large civil engineering or building projects, although for smaller projects, such as trench works for services, dewatering costs can rise to approximately 10 per cent of the main contract value. In the Roberts and Deed (1994) study over 80 per cent of groundwater control projects were valued at less than £50 000 (at 1994 costs) in the final account and the pumping period was less than 26 weeks.

Environmental constraints can have significant influence on groundwater control works. It is advisable to approach the appropriate environmental regulator (see Section 4.2) early in a project so that any relevant constraints can be identified. Ideally this should be done by the clients' representative at the planning stage so that constraints can be drawn to the attention of designers and contractors and, where necessary, be included in the contract documents.

There are significant risks associated with groundwater control works. These should be identified and contractual arrangements made for their allocation and management.

## 3.3.2 Contractual arrangements

Several forms of contract are used for the procurement of civil engineering and building projects. Various national and international bodies produce standard forms of contract (and subcontracts in some cases) including:

- ICE (Institution of Civil Engineers)
- NEC (New Engineering Contract)
- JCT (Joint Contracts Tribunal)
- IChemE (Institution of Chemical Engineers).

The main parties in these contracts are the client (or employer), the client's representative and the contractor. Payment can be based on a priced bill of quantities, 'contract milestones' or programme schedule, target cost, cost reimbursement, lump sum or other arrangement. Discussion of these procurement methods is beyond the scope of this report. Further information on the control of risk and forms of contract is provided by Potter (1995), Godfrey (1996) and Perry *et al* (1985). Information specific to tunnelling contracts is given by Attewell (1995). In many cases the client will appoint a client's representative (called the engineer under some forms of contract) to administer (and in some cases supervise) the works. The client's representative will arrange for a contractor to undertake the construction work. The contractor may appoint a specialist subcontractor to design, install and operate the groundwater control system.

Groundwater control works are typically put out to competitive or negotiated tender by the contractor under a standard form of subcontract. These are 'back-to-back' contracts so that the responsibilities and liabilities of the contractor are passed on to the subcontractor for that specific part of the works.

In practice it may be in the interests of all parties for the risks to be dealt with equitably in order to minimise disruption and control environmental and health and safety hazards. Potential risks ought to be clearly identified in the contract documents and realistically allocated between the employer, contractor and specialist subcontractor. Some matters to be considered when drawing up contract documents for groundwater control works are given in Table 3.3.

See also
7.2 Observational method

In addition to the conventional methods of subcontract procurement outlined above other arrangements are possible:

- For relatively straightforward schemes, particularly sump pumping and simple wellpointing operations, appropriate dewatering plant can generally be hired or even purchased. Where a contractor uses hired or purchased plant, the responsibility for the design and effectiveness of the scheme will generally remain with the contractor.

- A contractual arrangement, which promotes the use of the observational method (see Section 7.2) may be beneficial where the site investigation is not sufficiently comprehensive to confirm the design of a groundwater control scheme.
- Where the groundwater control works are long-term or integral to the design of the permanent works, it may be appropriate for the client to accept responsibility for the design and specification of the dewatering scheme.
- Systems of risk sharing can be developed. For example, an agreed minimum scheme can be specified with discounted rates for any additional equipment.

***Table 3.3 Some technical and administrative matters to be considered for groundwater control works***

| Subject | Examples |
|---|---|
| Specification of drawdown requirements | Artesian pressures in confined aquifers (eg Boxes 1.5 and 5.1)<br>Sensitivity of fine-grained soils to seepage pressures (eg Case study 7.3) |
| Achievement of drawdown | Drawdown requirements may be time-dependent or phased (eg Case study 7.10) |
| Programme | Mobilisation, installation and running periods need realistic assessment |
| Maintenance and security of drawdown | Potential for rapid recovery of groundwater levels (eg Box 3.4)<br>Responsibility for reacting to night-time or weekend breakdowns<br>Provision of standby plant and power supplies (see Section 3.4.4) |
| Monitoring arrangements and reporting | See Section 3.4 |
| Surface water | If not properly controlled, surface water can disrupt groundworks (see Section 2.1.1) |
| Discharge arrangements | Discharge permissions can take time to obtain and may include restrictions (see Section 4.2) |
| Off-site drawdowns | Differential settlement of buildings and buried services due to consolidation of compressible soil such as peat and soft silt and clay (eg Box 6.13 and Case study 7.8) |
| Environmental impact | Derogation of water supplies (eg Box 4.4)<br>Movement of contaminated groundwater (eg Box 4.2) |
| Access and headroom | Restrictions should be drawn to the attention of subcontractors |
| Buried services | Procedures for protecting and checking for services must be agreed |
| Assistance and attendance from main contractor, client or other subcontractors | Different dewatering techniques or subcontractors can require significantly different attendances |

## 3.3.3 Costs

**See also**
1.2.8 Dewatering costs

It is not possible to estimate groundwater control costs, even very approximately, from the quantity of water pumped, the volume of soil dewatered or the amount of drawdown achieved. As a consequence, no formal method of measurement has been developed for groundwater control works. The actual costs for dewatering works can generally be divided into two broad categories:

1. Method-related, eg mobilisation, installation, commissioning, demobilisation.
2. Time-related, eg plant provision, power or fuel supply, monitoring and supervision.

The time-related costs are generally significant, so the total costs for groundwater control works tend to reflect the duration of pumping (see Roberts and Deed, 1994).

## 3.3.4 Tender assessment

As with other specialist construction works, tender assessment of groundwater control schemes needs care, as:

- the technical merits of different schemes may not be clear to non-specialists
- subcontractors may request significantly different levels of attendances from the main

contractor, which can have a major impact on the apparent tender cost

- some proposals can offer improved access or flexibility that may reduce overall construction costs
- a subcontractor may have local knowledge, not available to others, which may allow them to offer a more competitive scheme or, conversely, a higher but more realistic price
- subcontract quotations may assume radically different contractual arrangements or risk management structures (eg hire arrangements can offer reduced costs but increased responsibility for the main contractor).

It is good practice for the subcontractor to provide a method statement which sets out the proposed scheme and defines the underlying design assumptions. Table 3.3 can be used as a check list of topics, which may have to be considered in the subcontract documents.

# 3.4 OPERATION AND MONITORING

## 3.4.1 Operation

**See also**
3.4.4 Standby facilities
3.4.5 Clogging and biofouling

Operation of a groundwater control system involves more than just switching the pumps on and starting to dig. Groundwater levels and system performance have to be monitored to make sure the specified performance targets will be, and are being, achieved, so that the excavation is maintained in a safe and stable condition. Maintenance of the pumping equipment is also necessary.

Nevertheless, monitoring should not be undertaken as a matter of course or because it seems the 'right' thing to do. The monitoring should be an integral part of the safety and quality management system on site. Merely taking the readings and filing them away is not sufficient; the results should be plotted in a way that highlights the performance of the system and be displayed for engineering and management staff. In addition, they should be regularly reviewed by a nominated member of the site management team, and any observed changes or trends in the data investigated, if necessary by obtaining specialist advice. In many cases the stability of the excavation is critically dependent on the groundwater control system. The use of a 'traffic light' system of green, amber and red trigger levels with agreed actions can be an efficient way of managing and presenting monitoring data.

The performance of a groundwater control system may deteriorate for a variety of reasons, including mechanical problems with pumps, clogging of wells or biofouling (see Section 3.4.5). Only by a programme of monitoring can these potential problems be recognised, and action taken, before a major problem develops.

The maintenance of the system depends on the equipment used. Diesel-powered plant (pumps or generators) will require fuelling and coolant or lubricant levels need to be checked and replenished in accordance with the manufacturer's or hirer's recommendations. Electrical pumps generally require less maintenance on site, but switchgear should be tested regularly in accordance with the Institution of Electrical Engineers Wiring Regulations (BS 7671:2008). Standby plant (Section 3.4.4) should be tested by running on load. Any alarms designed to signal system failure should also be regularly tested.

Monitoring and maintenance should be carried out by a nominated member of the site staff during normal working hours and at weekends. On large projects, or where the control of groundwater is critical to the stability of an excavation, an experienced dewatering site operator may be required in order to provide overnight emergency cover as well as to carry out the monitoring and maintenance. The operator may be resident near the site, with an automated telemetry and alarm system used (see Section 3.4.4) to alert the operator in the event of system failure during the night.

## 3.4.2 Monitoring and record keeping

**See also**
Box 3.5 Monitoring

The scale of the monitoring programme should correspond to the complexity of the groundwater control system and to the potential consequences of system failure. The monitoring requirements at different stages of a project are shown in Table 3.4.

*Table 3.4* ***Key requirements at each stage of a monitoring programme***

| Stage | Monitoring requirements |
|---|---|
| Pre start-up | Compare jetting records or well drillers' logs against site investigation data<br>Determine initial groundwater level<br>Determine reduced levels of monitoring points and datum<br>Carry out initial level survey and dilapidation survey of existing structures (if significant settlement is expected) |
| Start-up and commissioning | Check functioning of pumps and equipment<br>Measure flow rate and drawdown to check targets are met (system to be modified or adjusted if required)<br>Test groundwater quality to check conditions of discharge consent are satisfied<br>Check adequacy of power supply, discharge point and standby facilities<br>Carry out switch-off test to determine rate of recovery |
| Operation and running period | Establish monitoring regime (see Table 3.4)<br>Establish fuelling and plant maintenance regime<br>Monitor settlement and condition of structures (if significant settlements are expected)<br>Check regularly for damage to, or burial of, equipment |
| Switch-off and decommissioning | Monitor recovery of groundwater levels as pumps are switched off to check that stability or floatation problems do not occur<br>Pumps may need to be switched off sequentially over several days to avoid sudden rises in water levels |

Typical monitoring requirements for relatively simple projects are shown in Table 3.5. Monitoring during the start-up and commissioning period could be more frequent, but once the target drawdown has been achieved, the monitoring frequencies given in Table 3.5 would usually be appropriate. An example of a weekly monitoring record sheet is shown in Box 3.1.

*Table 3.5* ***Typical monitoring programme for the operational period of a simple groundwater control project (after Roberts and Preene, 1994b)***

| Parameter | Method | Frequency of monitoring |
|---|---|---|
| Mechanical performance | Vacuum (wellpoints)<br>Supply pressure (ejector wells)<br>Power supply alarms<br>Diesel engine checks | Daily |
| Standby equipment | Run standby pumps and generators on load | Daily or weekly[1] |
| Drawdown in observation wells | Measured by dipmeter or datalogger monitoring equipment, relative to a known datum | Daily |
| Flow rate, system total | Measured by V-notch weir, flowmeter or volumetric measurement | Daily |
| Discharge quality | Visual inspection of discharge tanks to check for suspended solids or oil contamination<br>Turbidity tube or turbidity meter used to check clarity of discharge[2]<br>Chemical testing of discharge water[2] | Daily<br><br>Weekly or monthly<br>Weekly or monthly |
| Drawdown in pumped wells | Measured by dipmeter or datalogger monitoring equipment, relative to a known datum | Daily, weekly or monthly |
| Settlement effects | Level surveys of selected points[2]<br>Check existing structures for signs of distress[2] | Weekly or monthly |

**Notes**

1 Depending on the rate at which groundwater levels recover.

2 May not be required for all projects.

Long-term trends in system performance, or any external effects, are easier to identify if the monitoring data are plotted in graphical form (Box 3.5). Deterioration of system performance occurs for a variety of reasons:

- chemical clogging or biofouling
- loss of pump performance from wear-and-tear
- obstructions in discharge tanks or pipework
- accidental damage to the system resulting from other site activities
- inadequate adjustment or maintenance of system.

External effects, which can affect performance, include:

- groundwater control operations on other nearby sites
- pumping from nearby water supply wells
- variation in levels of surface water in connection with the aquifer (eg tides)
- natural seasonal or climatic variations in groundwater level (eg during periods of unusually high or low rainfall).

***Box 3.1 Example of a weekly record sheet***

A weekly record sheet allows data taken on site to be clearly recorded. As well as discharge flow rates and groundwater levels, equipment performance, alterations and testing of standby equipment are noted.

**WJ GROUNDWATER LIMITED** Week commencing
**Site:** Operator
**Client:** **Job No.** Shift Day/Night

**DEEPWELL SYSTEM WEEKLY MONITORING RECORD**

| **Daily checks** | | Sat | Sun | Mon | Tue | Wed | Thu | Fri |
|---|---|---|---|---|---|---|---|---|
| Discharge tank 1: V-notch | mm | | | | | | | |
| tank flowrate | l/s | | | | | | | |
| Discharge tank 2: V-notch | mm | | | | | | | |
| tank flowrate | l/s | | | | | | | |
| discharge quality - visual | clear/cloudy | | | | | | | |
| number of pumps running | | | | | | | | |
| number of pumps off | | | | | | | | |
| Groundwater level monitoring | | | | | | | | |
| Time readings taken | | | | | | | | |
| Well No. 1 ( Datum | mOD) | | | | | | | |
| Well No. 2 ( Datum | mOD) | | | | | | | |
| Well No. 3 ( Datum | mOD) | | | | | | | |
| Well No. 4 ( Datum | mOD) | | | | | | | |
| Well No. 5 ( Datum | mOD) | | | | | | | |
| Well No. 6 ( Datum | mOD) | | | | | | | |
| Well No. 7 ( Datum | mOD) | | | | | | | |
| Well No. 8 ( Datum | mOD) | | | | | | | |
| Well No. 9 ( Datum | mOD) | | | | | | | |
| Duty generator running smoothly | check | | | | | | | |
| Diesel tank level | depth (cm) | | | | | | | |
| Cable routes clear | check | | | | | | | |
| Discharge pipe routes clear | check | | | | | | | |
| Damage risk to system | check | | | | | | | |

**Record of alterations/damage to system/maintenance carried out**

Sat:
Sun:
Mon:
Tue:
Wed:
Thu:
Fri:

Weekly checks

Standby generators tested Date: Time: Hours run:
Alarm tested Date: Time:

**List of wells running at the end of the week:**

Other comments:

A monitoring regime should specify criteria for when action has to be taken or modifications made to the system. The critical factor affecting safety and stability is usually the drawdown (ie the lowered groundwater level) within the excavation. Drawdown is typically monitored by recording groundwater levels (Box 3.2) in observation wells or piezometers (see Figure 5.2) with response zones in the appropriate aquifer. A set of trigger levels for the groundwater levels represents a suitable criterion: if water levels in observation wells or piezometers rise above the trigger level, remedial action is necessary. Monitoring of an ejector well project is illustrated

**See also**
Box 3.5 Monitoring
Figure 5.2 Piezometers

in Box 3.5. It is good practice to install datalogging monitoring equipment in at least one observation well to provide a continuous record of groundwater levels during a project.

*Box 3.2 Methods of measuring groundwater levels*

Groundwater levels are usually monitored in unpumped wells or observation standpipes or piezometers (see Section 5.2.2) with a dipmeter. Pore pressure transducers linked to electronic datalogging equipment may be used for automatic monitoring of water levels and to provide an alarm function. These should be installed, calibrated and recalibrated periodically in accordance with the manufacturer's instructions.

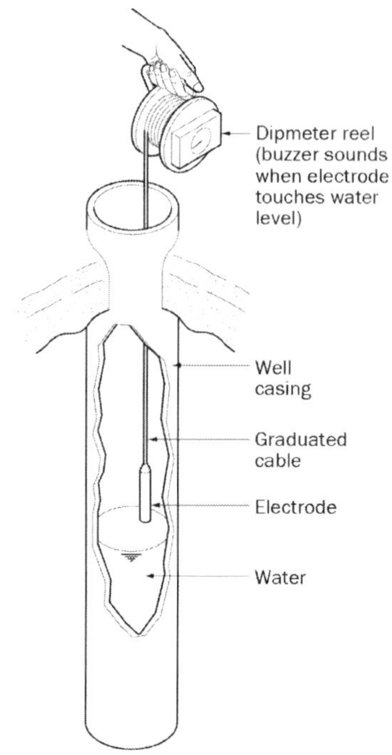


***Dipmeter for measuring depth to water in a well or piezometer***

## Methods of measuring discharge flow rate

Discharge is commonly measured by flowmeters, volumetric measurement or weir tanks. Two main types of meter are available:

- totalising meters, which record total volume of flow (average flow rate can be calculated from two readings at known time intervals)
- transient meters, which measure flow rate directly (some types can also record total flow).

Flowmeters should be installed into the discharge pipework in accordance with the manufacturer's instructions, including locating the meter away from valves and with adequate lengths of straight pipe provided on either side (normally a length of straight pipe of ten pipe diameters is required upstream and five diameters downstream). Flowmeters generally require full bore pipe flow free of entrained air. Flowmeters are susceptible to clogging by biofouling deposits and may require periodic recalibration and maintenance.

Volumetric determinations of low to moderate flow rates can be made by using a stopwatch to record the time taken to fill a container of known volume. At low flow rates (less than 10 l/s), provided a sufficiently large container is used (typically 40 to 200 litres), this can be a very accurate method.

**See also**
A1 Datasheet 3 V-Notch weir discharge charts

V-notch or rectangular notch weirs installed in settlement tanks connected to the discharge pipework can be used to estimate flow rate. The depth of water running over the weir is measured (Box 3.3) and a discharge chart is used to determine the flow rate (see Appendix A1, Datasheet 3).

*Box 3.3 Flow rate measurement by V-notch weir*

The depth of water, h, over the weir is measured above the base of the V-notch. The position of measurement should be upstream from the weir plate by a distance of approximately 0.1 m to 0.7 m, but not near a baffle or in the corner of a tank. Baffles may be required to smooth out any surges in the flow.

Datasheet 3 (Appendix A1) gives discharge charts for V-notches of $\alpha$ = 30°, 60° and 90°.

Specifications for weirs and tanks are given in BS ISO 1438:2008.

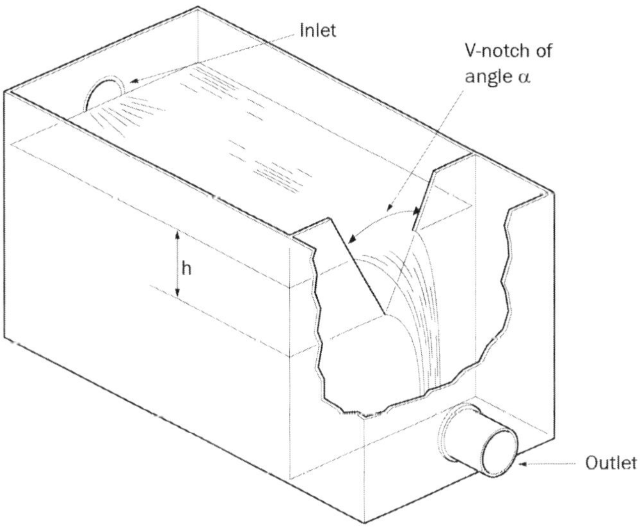


## 3.4.3 Discharge arrangements and monitoring

**See also**
4.1 Environmental impacts
4.2 Regulatory framework

Proper management of the discharged water is an essential part of any groundwater control scheme. Discharge permissions are required for all groundwater discharges (see Section 4.2.3). Disposal options for discharge water include:

1. **To surface waters (ie river, watercourse, lake, sea).** In England permission is required from the Environment Agency (EA), in Wales permission is required from Natural Resources Wales (NRW), in Scotland permission is required from the Scottish Environment Protection Agency (SEPA), and in Northern Ireland permission is required from the Northern Ireland Environment Agency (NIEA).
2. **To groundwater (ie via soakaways, recharge wells, or recharge trenches).** Consent is required from the EA, NRW, SEPA or NIEA.
3. **To an existing sewer.** Permission is required from the sewerage authority (eg water utilities or their agents), who may levy a charge for disposal of water in this way.

Discharge arrangements should minimise environmental impact (see Section 4.1). It is common practice to pass discharge flows through a weir tank (such as the one shown in Box 3.3) so that the flow rate and the clarity of the discharge water can be inspected. A cloudy discharge may indicate the presence of suspended solids in the water that might harm the aquatic environment (Section 4.1.1). If the discharge contains silt, a settlement lagoon may be needed. Another potential problem is the erosion of surface watercourses by poorly arranged discharges washing away river banks or beds. In many cases the use of protective slabs, mats or bales can prevent or minimise this problem. The use of lagoons and erosion protection measures applies not only to discharges from pumping, but also to water runoffs from wellpoint jetting (Section 2.1.3) or well development (Section 2.1.5), when sediment-laden water is often generated for short periods.

### Methods of determining discharge quality and chemistry

The discharge permission issued by the regulatory authority may prescribe constraint limits for the water chemistry and the suspended solids content of the discharge water. Water chemistry is usually measured by taking samples from the discharge at specified times for testing at an off-site laboratory – the discharge point should be accessible for sampling. Methods for obtaining and handling samples are given by Misstear *et al* (2006), Harris *et al* (1995), and BS ISO 5667-11:2009. Parameters to be tested are normally specified in the discharge permission. The clarity of the discharge water can be assessed using a turbidity meter or tube. The tube allows turbidity to be measured by determining the depth of water which, when viewed from above, just obscures the markings at the base of the tube.

If any type of oil (such as diesel fuel) is spilled on site or leaks from bowsers or plant, the oil may be drawn into the dewatering system and contaminate the discharge. The oil will appear as a coloured film on top of the water in discharge tanks or lagoons. Table 3.6 gives the amount of oil contained in films of various thicknesses. If spills occur, specialist advice should be obtained immediately and remedial measures taken (see Section 4.1.4).

*Table 3.6 Appearance of oil films on water (after Fussell et al, 1981)*

| Appearance of oil film on water | Approximate thickness µm | Approximate quantity of oil in film l/m² |
|---|---|---|
| Barely visible under the most favourable light conditions | 0.04 | $4.4 \times 10^{-5}$ |
| Visible as a silvery sheen | 0.08 | $8.8 \times 10^{-5}$ |
| First trace of colour observed | 0.15 | $1.8 \times 10^{-4}$ |
| Bright bands of colour | 0.3 | $3.5 \times 10^{-4}$ |
| Colours begin to turn dull | 1.0 | $1.2 \times 10^{-3}$ |
| Colours are much darker | 2.0 | $2.3 \times 10^{-3}$ |

## 3.4.4 Standby facilities

Standby facilities are essential for any groundwater control system where a breakdown or interruption of pumping will cause instability or flooding of the excavation. Only where groundwater levels recover very slowly, or if the rise in water levels will not cause problems, should standby facilities not be provided. For wellpoint and ejector systems, where each pump operates many wells, standby pumps are usual. For deepwell systems, where many pumps operate in concert, it is not usually necessary to have a standby for each pump – typically one or two submersible pumps will be held in store on site as replacements for any units which fail. Electrically powered systems (mains supply or duty generator) should have a standby generator as a back-up power supply.

Modern electronics enable groundwater control systems to be fitted with alarms that trigger in various conditions, including:

- failure of duty or standby power supply
- failure of individual pumps
- loss of vacuum (wellpoints) or supply pressure (ejector wells)
- water level in well or piezometer rising above specified level
- discharge flow rate falling below specified level.

Alarms should have a battery back-up so that they will function during a power failure. Alarm sensors can trigger flashing lights, sirens and telemetric equipment linked to radio and telephone pagers to signal an alarm condition. A rapid changeover from duty to standby facilities can be achieved by using an automatic mains failure (AMF) system with sequential pump starter, which can switch over the power supply and restart the pumps in less than one minute.

To assess the need for standby facilities, the consequences of the pumps being off and the rate at which water levels would recover can be estimated by carrying out a switch-off or recovery test when the groundwater control system is initially completed, but before excavation starts. A switch-off test is described in Box 3.4.

*Box 3.4* ***Case history of a switch-off test to estimate the rate of recovery of groundwater levels***

A system of deepwells was installed around an excavation 8.1 m deep underlain by a confined aquifer. The purpose of the well system was to reduce pore water pressures to prevent base heave. When the system was commissioned, target drawdowns were achieved. Because groundwater levels and pore water pressures often recover very rapidly in confined aquifers, a switch-off test was carried out before excavation commenced. The system was switched off for 15 minutes; water levels in the wells were monitored and rose by 4 m in the first four minutes. This indicated that if the power supply failed when the excavation was at full depth, recovery of water levels would create a risk of base heave within a few minutes, with major consequences to the works. In order to guard against this, the standby generator was fitted with an AMF system to restart the pumps. The system operated a siren (to warn workers to leave the excavation) and a radio pager to alert a resident site operator, who would check that the AMF system had functioned correctly. For the short period the excavation was at full depth and was most vulnerable, a second standby generator was installed in case there were problems with the primary standby.

## 3.4.5 Clogging and biofouling of wells and pipework

Groundwater control systems required to pump for prolonged periods of time (more than a few months) may become encrusted with chemical precipitates or covered with bacterial growth (biofouling) – clogging of well screens, pumps and pipework may follow. Encrustation and biofouling stem from the natural chemical compounds and bacteria contained in groundwater. The biofouling process is explained in more detail in Howsam (1990) and in Howsam *et al* (1995). In saline or brackish groundwaters, the groundwater chemistry may promote corrosion of pumps and equipment.

### Chemical encrustation

Groundwater naturally contains chemical compounds in solution. When groundwater flows into a well, it undergoes a fall in pressure, and possibly aeration. This can lead to the precipitation of insoluble chemical compounds, which build up as scale deposits on well screens and pumps. The deposits may be iron or manganese oxides or carbonates or, especially in chalk or limestone aquifers, calcium carbonates. Unless these scale build-ups are severe, they are unlikely to affect operation significantly. Powers *et al* (2007) indicates the possibility of troublesome encrustation where the water hardness is greater than 200 mg/l of $CaCO_3$. Powrie and Roberts (1995) describe a site where several pumps became severely clogged by calcium carbonate build-up that was probably initiated artificially by the addition of free lime to the groundwater resulting from poorly controlled underwater concrete placement.

### Biofouling

**See also**
3.4.2 Monitoring

Most shallow groundwater is naturally teeming with micro-organisms; wells and pumping equipment may offer an environment in which these bacteria can thrive. The residue from the bacterial growth can lead to troublesome encrustation of wells screens, pumps and pipework. The process is known as biofouling. The most common form is the build-up of a soft red-brown gelatinous slime (biomass), which results from the action of iron-related bacteria such as Gallionella or Crenothrix. These aerobic organisms use oxygen from their environment to transform dissolved iron in the groundwater from a soluble to an insoluble state. The resulting iron oxides and oxyhydroxides combine with the slime produced by the bacteria to form a much greater volume of encrustation than would otherwise occur. The biofouling encrustation can be tenacious. After a pump is removed from a well, the deposits are soft and can simply be wiped off, but in the well the biomass will not be dislodged even by the highest groundwater velocities usually generated. If not cleaned by other means, the biomass will build up and may totally clog wells, pumps and pipework (Figure 3.2).

Any system pumping groundwater for prolonged periods may be at risk from biofouling. The results of a monitoring scheme measuring both discharge flow rate and drawdown (Section 3.4.2) will show whether the wells and equipment need to be cleaned or rehabilitated. As biofouling deposits build up, the discharge flow rate will decrease; if no action is taken, groundwater levels may rise to a point where instability or flooding occurs. A programme of well cleaning should prevent this (Box 3.5).

*Box 3.5 Case history of monitoring of drawdown for ejector well project where biofouling occurred*

Groundwater levels were monitored daily in a series of observation wells within a large excavation enclosed by an ejector well system. After the first few months of pumping, the groundwater levels rose gradually (shown below), and the discharge flow rate decreased from 5.5 to less than 3 l/s. The rise in groundwater level is characteristic of clogging of wells and equipment by biofouling. When trigger groundwater levels were approached, the wells were cleaned; groundwater levels fell immediately to close to their original levels. Over the next few months the wells were cleaned when trigger levels were approached. However, monitoring showed that each successive cleaning was less effective than the last. Once this was identified, a plan was developed to replace key ejector components, which overcame the decrease in the effectiveness of cleaning.

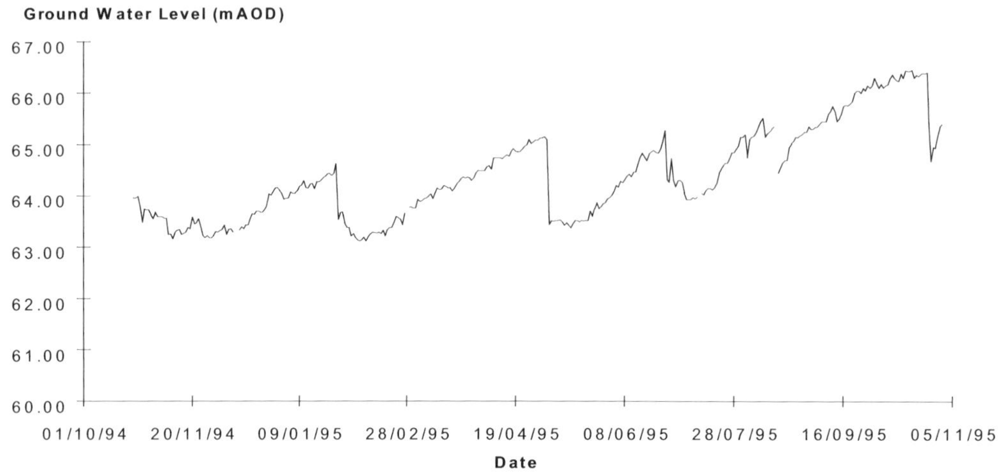


*Groundwater level monitoring for ejector well system*

Before pumping, chemical testing of groundwater samples may indicate the risk of clogging from biofouling (see Table 3.7). The likelihood of biofouling is not simple to predict, but is related to the concentration of dissolved iron in the groundwater, the flow rate and the type of system in use.

*a Reduction in pipework internal diameter due to biofouling*

*b Biofouling deposits on a borehole submersible pump*

*Figure 3.2 Encrustation of dewatering equipment due to biofouling (courtesy Geoquip Water Solutions Limited)*

The risk of biofouling increases with iron concentration in the groundwater, and with groundwater flow rates, as high flows provide the bacteria with a larger supply of oxygen and nutrients, allowing rapid growth. The type of dewatering system affects the risk of biofouling because the bacteria require an aerobic environment to thrive, so a wellpoint system (where most of the pipework is under vacuum) is far less susceptible to biofouling than wells with submersible pumps, where the water may be aerated as it enters the well. With ejector wells, clogging by biofouling is a problem because the recirculating water may concentrate loosened biomass and block the small passages in the ejector body. This can be avoided by a regime of regular cleaning. Recharge wells (see Section 2.3.3) are most prone to clogging, simply because any suspended matter in the recharge water will collect in the wells. A recharge system should be designed so that the water is aerated as little as possible, in order to retard biofouling, otherwise biofouling may be so severe that treatment and filtration is required, or in extreme cases recharge using the abstracted water may not be viable.

*Table 3.7 Tentative trigger levels for susceptibility to Gallionella biofouling (after Powrie et al, 1990)*

| Pumping technique | Susceptibility to biofouling | Concentration of iron in groundwater mg/l | Frequency of cleaning |
|---|---|---|---|
| Wellpoints | Low | <10<br>>10 | Biofouling unlikely to present difficulties under normal operating conditions and times of less than 12 months<br>Biofouling may be a problem for long-term systems |
| Submersible pumps | Moderate | <5<br>5–10<br>>10 | 6 to 12 months<br>0.5 to 1 month<br>Weekly (system may not be viable) |
| Ejector wells (low flow rate, <10 l/minute) | Moderate | <5<br>5 to 10<br>10 to 15 | 6 to 12 months<br>Monthly<br>Weekly (system may not be viable) |
| Ejector wells (high flow rate, >20 l/minute) | High | <2<br>2 to 5<br>5 to 10 | 6 to 12 months<br>Monthly<br>Weekly (system may not be viable) |
| Recharge wells | Very high | Recharge wells are prone to biofouling, which is likely to occur even if iron concentrations are below 0.5 mg/l.<br>To minimise biomass growth and encrustation extreme care should be taken to avoid aerating the recharge water.<br>It is not uncommon for recharge wells to require cleaning on a weekly or monthly basis.<br>Recharge wells using abstracted groundwater may require treatment or filtration and may not be viable at high iron concentrations. | |

**See also**
Box 2.3 Well development
Box 2.8 Recharge wells
Chapter 4 Environmental matters

When biofouling occurs, wells and equipment can be cleaned in several ways:

- Flushing the well with compressed air to loosen and pump out the biomass. This method is especially suited to ejectors and recharge wells (see Box 2.8). No craneage is needed because risers do not have to be removed from the well, and the well is only out of commission during cleaning.
- Removing the pumps and risers from the wells and cleaning them at ground level by jet washing or scrubbing. Submersible pumps may have to be disassembled to clean internal components. Airlift or jetting development methods (Box 2.3) can be used to clean out the well liner and screen.
- Chemical treatment of water in wells and pipework to break down the biofouling, which is then flushed away. Specialist advice should be sought, and due care taken in the handling of chemicals, and in the disposal of effluents.

Effluents of discoloured and sediment-laden water may be produced by all of these cleaning methods. These must be disposed of in accordance with the appropriate environmental legislation (see Chapter 4).

## Corrosion

Saline or brackish waters (indicative of significant levels of chloride) can lead to very severe corrosion problems, including the corrosion of stainless steel. In the UK groundwater may be saline or brackish in estuarine or coastal areas, where seawater has intruded into the aquifer. In arid climates, such as the Middle East, groundwater may be naturally of very high salinity as a result of the high levels or evaporation and low levels of recharge to aquifers.

The mechanisms and chemistry of groundwater related corrosion are complex and include:

1 Microbially induced corrosion. Stainless steels may become susceptible to attack by chloride ions because the passive oxide layer, which normally prevents corrosion, cannot re-form in the anaerobic environment created beneath a thick biofilm of bacteria. Even though a borehole environment may be aerobic, the growth of Gallionella bacteria on a metal surface

may generate anaerobic conditions beneath the biofilm, as the Gallionella use up all the available oxygen at the outer surface. If sulphates are present in the groundwater, the anaerobic conditions can allow sulphate-reducing bacteria to produce sulphides, which can form sulphuric acid that is very corrosive to cast iron and steel. Regular cleaning to remove the Gallionella biofilm will help slow down corrosion.

2 Electro-chemical corrosion. This can occur where a piece of equipment is made of more than one type of metal in contact with each other. The more susceptible metal corrodes in preference to the other; cases have occurred where pump impellers have corroded away to nothing while the rest of the pump was untouched.

If corrosive groundwaters are expected, as much pipework as possible should be made from plastic. Pumps can be constructed from grade 316 stainless steel rather than the less resistant grade 304. Cathodic protection can also be used.

### Other problems

Algae can grow by photosynthesis in slow-flowing water such as that in sumps or open-topped discharge tanks, but rarely grow in the dark environment of wells and pipework. Recharge and ejector systems are prone to disruption by algae, which can be drawn around the system and clog pumps. Growth of algae can be avoided by using closed tanks or covering open tanks with opaque material to block out the sunlight.

Recharge wells (Section 2.3.3) are very susceptible to clogging unless the water is absolutely clear. In reality this is never the case – the recharge water will always contain some entrained air or gas bubbles, fine soil particles, chemical precipitates and biofouling or algae residues. The quality of the recharge water may be improved by filtration or chemical treatment; specialist advice should be obtained in such cases. Recharge wells (Section 2.3.3) can be cleaned by airlifting and pumping to remove the clogging matter.

## 3.4.6 Capping and sealing of wells on completion

On completion of the groundwater control works, after the pumping equipment has been removed, plastic well liners are normally left in place (steel well screens and liners are sometimes pulled out to be reused). In some circumstances, such as simple shallow wellpoint systems, the wells may be left unsealed, with the wellpoint riser just cut off below surface reinstatement level. Deeper wells, especially those penetrating aquifers used for public water supplies or those penetrating more than one aquifer, may need to be specially sealed. The purpose of the sealing is to block any vertical seepage paths which could allow contaminants to reach the groundwater. Sealing details will have to be agreed with the environmental regulatory authorities and, in some circumstances, wells may have to be completely backfilled with grout. Guidance is given in Environment Agency (2012).

# 3.5 KEY REFERENCES


CASHMAN, P M and PREENE, M (2012) *Groundwater lowering in construction: a practical guide to dewatering, second edition*, CRC Press, Boca Raton, USA (ISBN: 978-0-41921-110-5)

GODFREY, P S (1996) *Control of risk: a guide to the management of risk from construction*, SP125, CIRIA, London (ISBN: 978-0-86017-441-7). Go to **www.ciria.org**

HOWSAM, P (ed) (1990) *Microbiology in civil engineering: international proceedings of the Federation of European Microbiologial Societies symposium held at Cranfield Institute of Technology, UK*, 3–5 September 1990, E & F Spon, UK (ISBN: 0-41916-730-7), FEMS Symposium), Spon, London

HSE (2006) *Health and safety in construction, third edition*, HSG150, Health and Safety Executive, London (ISBN: 978-0-71766-182-2). Go to: **http://tinyurl.com/q5fdm6n**

HSE (2015) *Managing health and safety in construction: Construction (Design and Management Regulations) 2015. Guidance on Regulations*, L153, Health and Safety Executive, London (ISBN: 978-0-71766-626-3). Go to: **www.hse.gov.uk/pubns/priced/l153.pdf**

OVE ARUP & PARTNERS and GILBERTSON, A (2015) *CDM2015 – construction work sector guidance for designers, fourth edition*, C755, CIRIA, London (ISBN: 978-0-86017-756-2). Go to: **www.ciria.org**

POTTER, M (1995) *Planning to build? A practical introduction to the construction process*, SP113, CIRIA, London (ISBN: 978-0-86017-433-2). Go to: **www.ciria.org**

POWERS, J P, CORWIN, A B, SCHMALL, P C and KAECK, W E (2007) *Construction dewatering and groundwater control: new methods and applications, third edition*, Wiley, New York, USA (ISBN: 978-0-47147-943-7)

POWRIE, W, ROBERTS, T O L and JEFFERIS, S A (1990) "Biofouling of site dewatering systems". In: P Howsam, P (ed) *Microbiology in civil engineering: international proceedings of the Federation of European Microbiologial Societies symposium held at Cranfield Institute of Technology, Abingdon, UK*, Routledge, *UK*, 3–5 September 1990, E & F Spon, UK (ISBN: 0-41916-730-7), pp 341–352

ROBERTS, T O L and DEED, M E R (1994) "Cost overruns in construction dewatering". In: B O Skipp (ed) *Risk and reliability in ground engineering*, Thomas Telford Publishing, London (ISBN: 978-0-72771-986-7)

# 4 Environmental matters

This chapter is divided into two parts. The first part describes how poorly managed groundwater control works may lead to adverse environmental impacts and pollution, and briefly discusses possible mitigation measures. The second part outlines the legislative framework in place in the UK in January 2014 (legislation may change and the reader should check the current regulations).

It is important to realise that groundwater control works have the potential to cause adverse impacts and pollution. Where potential impacts are of concern advice should be obtained from suitably qualified and experienced specialists. It is also advisable to engage in early consultation with the environmental regulators.

## 4.1 POTENTIAL ENVIRONMENTAL IMPACTS OF GROUNDWATER CONTROL WORKS

Groundwater control operations have the potential to cause impacts on groundwater and also on the wider environment. In many cases these potential impacts can be mitigated through good design and practice.

Potential groundwater impacts are discussed in Preene and Brassington (2003). Advice on good design and site practice to avoid water pollution from construction projects is given in Masters-Williams *et al* (2001) and Murnane *et al* (2006).

Table 4.1 summarises potential environmental problems, which may be associated with groundwater control activities, together with some possible mitigation measures.

*Table 4.1 Examples of environmental problems and mitigation measures*

| Potential environmental problem | Mitigation measures |
|---|---|
| Suspended solids (silt-laden water)<br>(Section 4.1.1) | • choice of appropriate dewatering method<br>• correct filter design, selection and placement<br>• settlement tanks, settlement lagoons or lamellae tanks for removal of suspended solids |
| Scouring or erosion of watercourses<br>(Section 4.1.2) | • attenuation of flow using settlement lagoons<br>• scour protection (eg concrete slabs or straw bales) |
| Discharge water quality<br>(Section 4.1.3) | • pH adjustment (eg adding lime)<br>• water treatment systems |
| Pollution caused by construction works<br>(Section 4.1.4) | • provision of bunding around pumps, generators, bowsers and fuel storage and handling areas<br>• oil traps and separators |
| Contaminated land and existing site contamination<br>(Section 4.1.5) | • barrier walls (eg sheet-piles, HDPE or slurry walls)<br>• hydraulic controls (eg pump and treat systems)<br>• water treatment systems |
| Ground settlement<br>(Section 4.1.6) | • suitable groundwater control scheme to minimise drawdown or pore water pressure reductions in compressible strata<br>• artificial recharge of groundwater |
| Barriers to groundwater flow<br>(Section 4.1.7) | • suitable groundwater control scheme to minimise potential barrier effects |

continued...

...continued from

| Potential environmental problem | Mitigation measures |
|---|---|
| Pathways for groundwater flow (Section 4.1.8) | • design of wells and boreholes to avoid creation of groundwater pathways and to prevent artificial linkages between aquifers |
| Derogation of water supply wells (Section 4.1.9) | • monitoring of affected locations and, if necessary, replacement or augmentation of affected water supplies |
| Impact on groundwater dependent features (Section 4.1.10) | • suitable groundwater control scheme to minimise impacts on groundwater dependent features<br>• artificial recharge of groundwater |
| Saline intrusion (Section 4.1.11) | • artificial recharge of groundwater |
| Noise | • use of silenced diesel equipment or electrically powered pumps |

## 4.1.1 Suspended solids: silt

By far the most common instance of pollution from groundwater control operations is suspended solids in the form of silt, which, if discharged, causes harm to the aquatic environment (Box 4.1).

***Box 4.1 Harmful effects of silt on the aquatic environment***

Silt-laden water can:

- injure fish by its abrasive action
- clog the gills of fish, causing them to die by suffocation
- destroy spawning sites and insect habitats on the river bed, removing the source of food for fish
- reduce light levels underwater, affecting plant growth
- coat the leaves of aquatic plants, limiting their growth.

Silt discharges are unsightly and will be reported by the public.

**See also**
6.3.3 Filter design

The best way to manage suspended solids in discharges is to tackle the cause of the problem and design and specify the groundwater control system with adequate filters (see Section 6.3.3) to minimise sediment in the discharge water. Provided that suitable well screens and filters are installed, wellpoint, deepwell, suction well and ejector well systems do not normally produce discharges with high sediment contents, except during the initial periods of pumping and development, when dirty water may be produced for short periods. Pumping from wells in certain aquifers, for example Chalk, can give rise to very fine suspended sediments in the pumped water that can be hard to remove by conventional methods.

**See also**
2.1.2 Sump pumping

The method, which most commonly produces sediment-laden water, is sump pumping (Section 2.1.2). Installation of adequate filters around sumps can be difficult and, as a result, clay, silt and sand-size particles can be drawn to the pump and entrained in the discharge water. Whenever sump pumping is carried out, arrangements should be made, before final discharge, to remove any suspended solids to below the maximum levels set by the environmental regulator in the discharge permission. If this is not possible, it may be necessary to change to another groundwater control method, such as wellpoints, with adequate filters.

**See also**
Box 3.3 Settlement tank

Sand-size particles can usually be removed by passing the discharge through a grit trap or settlement tank. Such tanks typically have a minimum size of 3 m by 1.5 m and are approximately 1.5 m deep (see Box 3.3). The sand will build up in the base of the tank and will need to be cleared out periodically to ensure the tank continues to operate efficiently.

Silt and clay-size particles will not settle out naturally in small tanks. If silt needs to be settled out, large lagoons are sometimes used. These may require large areas of land (planning permission from the local authorities may be necessary) and typically consist of earthwork bunds around the perimeter with some form of waterproof lining on the base and sides. Lagoons will require edge protection to reduce hazards to personnel and should be designed to allow removal of sediment as necessary.

In recent years some innovative approaches have been developed to allow more effective settlement of solids via small tanks. One example is the lamella plate tank (Figure 4.1). In this system the water is passed upward between inclined parallel plates, which are separated by a small gap. Solid particles settle onto the inclined plates and fall to the base of the tank. The lamella plate approach allows a small tank, comparable in plan size to a conventional steel settlement tank, to have the settling capacity of a lagoon several times its size.

***Figure 4.1 Specialist lamella plate settlement tank (courtesy Cornelsen Limited)***

Where removal of suspended solids is necessary down to very low levels (eg if the receiving watercourse is especially sensitive), and if large lagoons are not feasible, it may be possible to increase the rate of sedimentation by chemical addition using flocculants (see Nyer, 2009). Chemical treatment of discharges is not a straightforward matter, and may require environmental permissions from the regulator – specialist advice should be sought.

### 4.1.2 Watercourses

The problem of suspended solids is made worse if there is erosion of the streambed or riverbank by uncontrolled discharge. Furthermore, the erosion or scouring action can cause long-term damage to the watercourse itself. In many cases correct management of the discharge can prevent or minimise the problem. Materials such as geotextile membranes, gabion baskets, stone mats or even straw bales can be placed at the discharge point to dissipate the energy of the discharge and reduce potential erosion.

Consent or permission from the environmental regulators or other bodies controlling the watercourse may be required for works in, over or under a watercourse (this may apply to erosion control measures). Works that may require consent include:

- alterations to the shape and alignment of a watercourse
- structures constructed within a watercourse, including discharge headwalls, bridges and coffer dams
- construction adjacent to watercourse, whether it affects the watercourse or not.

Given the historic and current focus on flood management in the UK, it is important to realise that any effect on or interaction with watercourses will be closely scrutinised by environmental regulators. Typically, regulators will not approve or consent to work that is assessed to harm the environment or increase flood risk, even if the works are structurally sound. They will discourage culverting, diverting or channelling watercourses and building into watercourses, known as encroachment.

### 4.1.3 Discharge water quality

Even where water from a groundwater control system is discharged in a controlled manner, with low suspended solids, if the receiving water environment is ecologically sensitive it will be necessary to assess the impact of the differences between the water quality of the discharge and receiving water.

Possible issues to be considered include:

- **Contamination of discharge water.** This can occur when dewatering is carried out on or near contaminated sites or if there is a spill or leak of fuel or chemicals on the construction site (see Sections 4.1.4 and 4.1.5).
- **Dissolved substances in discharge water.** Problems can occasionally occur where dissolved substances (eg iron or carbonates) present in the discharge water precipitate out in the discharge flow. This can result in unsightly deposits and possible adverse impacts on surface water quality and aquatic ecology. A common example is red-brown discolouration of the surface water and deposits on the streambed and plants where water is discharged with high concentrations of dissolved iron.
- **pH and salinity of discharge water.** If the pH and salinity of the discharge water is significantly different to that in the receiving surface water body, then there is the potential for adverse impacts on surface water quality and aquatic ecology.
- **Temperature of discharge water.** Groundwater temperature tends to vary little during the year, typically being close to the mean air temperature at a site (in the UK typical discharge water temperatures from dewatering systems are in the range 8°C to 14°C depending on site location and depth of well screens). Conversely, surface water temperature will tend to be more affected by seasonal air temperature variations. Therefore, in winter the discharge from a dewatering system will tend to be at a warmer temperature than surface water and in the summer the discharge will tend to be cooler than surface water. If dewatering water is discharged into a surface water environment that contains ecological habitats (eg plants or aquatic wildlife and especially salmonid rivers) that are sensitive to temperature, there is a risk of adverse impacts on local ecology.

Where the flow from a dewatering system is to be discharged to a potentially sensitive location, specialist advice should be obtained from suitably experienced hydrologists, ecologists and water quality specialists, to will ensure the necessary environmental permissions are obtained and that environmental impacts can be managed appropriately.

### 4.1.4 Pollution caused by construction works

**See also**
Table 3.7 Oil pollution

Construction activities in general, as well as those directly associated with dewatering and groundwater control works, have the potential to cause pollution. It is essential that good practice be adopted in construction to reduce the risk of pollution, and to deal promptly and effectively with any incidents that do occur. Guidelines on pollution prevention (Pollution Prevention Guidance notes, PPGs) have been produced jointly by the UK environmental regulators, to advise industry and the public on legal responsibilities and good environmental practice. These are available via the environmental regulators' websites (listed at the end of Section 4.2).

A particular risk with pumped dewatering systems is that oil or other petroleum products may be drawn into dewatering systems as a result of spills or leaks from plant (such as pumps and generators), fuel bowsers or tanks. Spills may occur during fuel deliveries, plant maintenance, or as a result of vandalism. To reduce the risk of leaks, fuel should be stored in secure bunded tanks and areas and operatives should be trained in correct handling procedures. Petroleum products may also be present in discharges when pumping on or near contaminated sites (see Section 4.1.5). Similar issues apply to any chemicals stored on site for use in construction.

Petroleum-based products are generally lighter than and mix poorly with water. These are known as light non-aqueous phase liquids (LNAPLs). LNAPLs will appear as a coloured film on the water surface in tanks, lagoons or watercourses. Table 3.6 gives an indication of the amount of oil contained in films of various appearances. If the discharge is oil-contaminated, oil traps or interceptors can be incorporated in the pipework to collect the LNAPL for periodic removal. Advice should be sought from the manufacturers of such equipment before it is employed. If oil films appear in tanks or lagoons, the techniques used for oil spills may be appropriate. These are described by Fussell *et al* (1981), and methods include the use of floating sorbent booms and pillows to draw the oil from the water surface, or the use of floating pumps to skim the oil layer off the water.

## 4.1.5 Contaminated land and existing site contamination

For all groundwater control systems the discharge of water will require some form of permission from the environmental regulator (or from the sewerage utility where water is discharged to sewer). Permissions are required even if the water being discharged is 'clean' and uncontaminated. However, there are obvious additional risks and concerns if the discharged water is contaminated in some way. In these cases further site investigations, water treatment measures and environmental permissions may be needed.

In the UK, it is not unusual for groundwater to be contaminated to some degree. The most common source of groundwater contamination is from the legacy of past industrial use of a site. It is important to note that some forms of groundwater contamination are relatively mobile and can migrate away from the original site under natural hydraulic gradients. This 'plume' of groundwater contamination can potentially migrate beneath neighbouring sites. Occasionally groundwater contamination can be present due to non-industrial activities such as leaking sewers or sewage treatment works, or from agricultural uses (eg pesticides and nitrates). There are several guides to the types of pollution that can be expected at industrial sites (for example Harris *et al*, 1995, DoE, 1987, Aspinwall and Co, 1994). The identification of potential groundwater contamination should be a key task at the desk study stage of site investigation (see Section 5.2).

If groundwater is pumped as part of a groundwater control scheme it is possible that groundwater contamination will be mobilised and will be drawn to the pumping system, and will emerge in the pumped water. If the source of contamination is on or close to the site where pumping is being carried out, contamination may be apparent in the discharge water as soon as pumping begins. If the source of the contamination is further away, but within the ultimate zone of influence of the dewatering system, the contamination may be drawn slowly toward the pumping system and may only be apparent in the discharge water after weeks or even months of continuous pumping. If significant contamination is anticipated in the pumped discharge water, it is normally necessary to install a temporary groundwater treatment plant on site to reduce the concentrations of contamination in the pumped water to acceptable levels (ie below the levels specified by the environmental regulator) before discharge.

Where groundwater contamination is considered to be a potential issue, advice should be obtained from suitably qualified and experienced specialists in contamination investigations and assessment, water treatment and environmental permitting. It is also advisable to engage in early consultation with the environmental regulators.

Contamination from organic pollutants (ie hydrocarbons) and, in particular, LNAPLs presents one of the most common forms of industrial pollution. This is because of the high mobility of the liquids, which are lighter than water and consequently float on the surface of the groundwater table. In addition, although LNAPLs are relatively insoluble, the low solubility levels can still exceed groundwater quality standards and discharge permissions limits.

Treatment technologies for contaminated groundwater are constantly being developed and evaluated. A guide to the range of technologies available for different compounds of organic pollutants and heavy metals is presented in Table 4.2. Further details can be found in Nyer (2009), Harris *et al* (1995) and Holden *et al* (1998).

On projects where one of the objectives is to remediate or 'clean up' contaminated land, the control of groundwater can be an integral part of the remediation strategy. Groundwater control methods can be used as part of an *in situ* treatment technique, eg pump and treat methods (Holden *et al*, 1998). Alternatively, for pollution at shallow depths, groundwater control may be needed to provide a stable excavation from which to remove and treat the contamination *ex situ*. An example of the use of sump pumping as part of a remediation strategy is given in Box 4.2.

Specific issues that need to be considered in the design, installation, operation and decommissioning of groundwater control systems on contaminated sites include:

- Sites where ground gas (for example methane, carbon dioxide and hydrogen sulphide) may be present will need additional health and safety measures. This may include access, excavation and storage locations being designated as confined spaces because of the potential for severe injury through oxygen depletion or explosion. There may be the requirement for explosion proof and intrinsically safe equipment, with a corresponding increase in costs and complexity of operation.
- Excavation or drilling into contaminated soils or rocks or pumping of contaminated groundwater could become a human health risk through direct exposure of the workforce or the public through breathing, ingesting or contact with skin.

Where work is carried out on or near potentially contaminated sites, advice should be obtained from suitably qualified and experienced health and safety specialists.

*Table 4.2 Technologies for treating contaminated groundwater (after Holden et al, 1998)*

| Treatment stage | Treatment methods |
|---|---|
| Preliminary treatments[1] (if required) | • equalisation of mixing water from different wells<br>• oil-water separation<br>• pH adjustment<br>• dilution if contaminant concentrations exceed operating range for treatment processes used<br>• prechlorination to minimise biological fouling |
| Primary treatments (if required) | • solids removal<br>• coagulation and flocculation followed by sedimentation to remove metals or solids |
| Secondary treatments (main treatment) | • biological treatment<br>• chemical precipitation or reduction<br>• air stripping<br>• chemical oxidation<br>• carbon adsorption<br>• membrane systems |
| Tertiary treatments | • biological treatment<br>• chemical oxidation<br>• carbon adsorption<br>• membrane systems<br>• UV oxidation<br>• rapid gravity sand filtration<br>• ion exchange |

**Note**

1 Removal of suspended solids by sedimentation should not be necessary if the extraction wells are properly developed

*Box 4.2 Case study of contaminated land remediation involving groundwater control*

**Background:** a backfilled quarry adjacent to a canal was contaminated with heavy hydrocarbons from a former bitumen works. In order to develop the site for future industrial use, the degree of contamination had to be reduced to a standard agreed with the development authority.

**Ground conditions:** the site was originally overlain with completely weathered Sherwood Sandstone, which had been excavated and replaced with a backfill of sand tailings from the quarry operation. An old river channel containing river terrace gravels passed through the site. The groundwater levels were predictably high at the site.

**Remediation strategy:** a clay cut-off wall was constructed around the site to contain the volume of soil and control peripheral groundwater levels. The clay cut-off wall was installed into underlying alluvial clays. Sump pumping was used to remove the contaminated groundwater and stabilise the excavation in the backfilled sand.

**Treatment system:** the pumped water was stored in a settlement lagoon. LNAPLs were skimmed off the surface of the lagoon for disposal. The remaining water was treated using an activated carbon filter to a quality acceptable for discharging to the ground via recharge trenches.

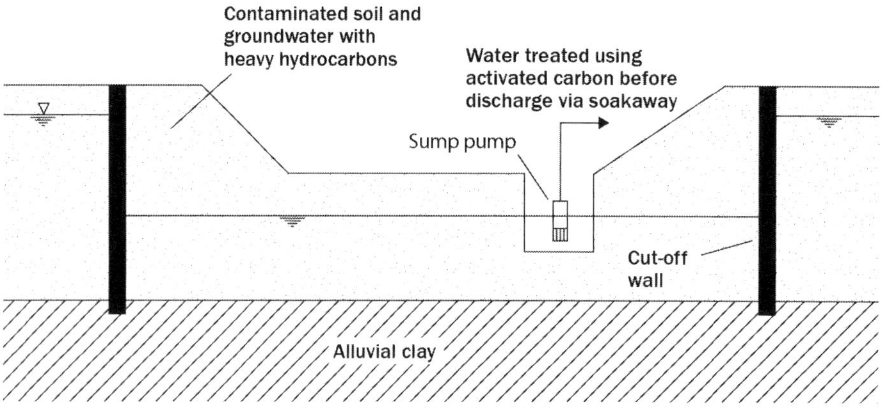


**Excavation and treatment of contaminated backfilled quarry**

## 4.1.6 Ground settlement

**See also**
2.3 Recharge systems
6.6 Settlement

Some ground settlement will occur every time groundwater levels are lowered by groundwater control operations. In the vast majority of cases the ground settlements will be so small that no distortion or damage is apparent in nearby structures. However, occasionally settlements may be large enough to cause potential problems.

Groundwater control operations can generate ground settlement by three principal mechanisms:

1. Increases in effective stresses due to lowering of groundwater levels. Lowering of groundwater levels naturally results in an increase in effective stress and a corresponding compression of soil and rock layers. This compression creates the potential for ground settlements across the zone of influence of the dewatering system.
2. Loss of fine particles from the soil. If wells and sumps do not have filters, which prevent the continuous removal of fine particles with the pumped water, then the removal of particles will loosen the soil and possibly lead to the creation of underground erosion channels or 'pipes'. This can cause localised settlement.
3. Instability of excavations when groundwater is not adequately controlled. If the groundwater lowering or pore water pressure reduction is inadequate (see Case study 7.6), then the resulting instability may lead to collapse of the excavation and significant localised settlement.

In most cases good design and practice can prevent settlement due to loss of fines and settlement due to instability. However, effective stress settlements will occur on all groundwater control projects where groundwater levels are lowered. The magnitude of the settlements will depend on a number of factors:

- The presence and thickness of any compressible layers of soil below the groundwater level, which will be affected by the pore water pressure reduction. Examples include soft alluvial silts and clays or peat deposits. The softer a soil layer (and the thicker it is), the greater the potential settlement.

- The amount of drawdown. The greater the drawdown of the groundwater level, the greater the resulting settlement.
- The period of pumping. Where the compressible soil is of low permeability (eg silts and clays) then the consolidation will be time dependent so that the longer pumping is continued, the greater the settlement.

Methods for assessing effective stress settlements due to groundwater lowering are described in Section 6.6. Further background is given in Preene (2000).

Where ground settlements are assessed to be large enough to potentially damage buildings, structures or buried services then some form of mitigation may be required. Artificial recharge of groundwater (Section 2.3) is often used to reduce the drawdown of groundwater around vulnerable structures, thereby reducing the ground settlements and risk of damage (see Case studies 7.9 and 7.11).

### 4.1.7 Barriers to groundwater flow

Where groundwater control measures include physical cut-off walls (see Table 1.2) these will form low permeability barriers to exclude horizontal groundwater from excavations. For most of the commonly used techniques, the resulting groundwater barrier is effectively permanent and will remain in place following the end of the construction period, and may interrupt horizontal groundwater flow, causing a damming effect and altering groundwater levels local to the structure. Typically these effects are not significant unless large cut-off structures (such as long linear structures for metro stations or cuttings for roads or railways) fully penetrate significant aquifer horizons. For such cases the potential rise of groundwater levels upgradient (and the corresponding lowered groundwater levels downgradient) of the structure should be considered during design and in the planning of groundwater monitoring systems.

Where impacts from groundwater barriers are a potential concern then numerical groundwater modelling can be used to investigate the changes in groundwater level. If impacts are considered to be significant then consideration should be given to modifying the cut-off wall, or to use cut-off walls, which are temporary in nature (such as steel sheet-piles removed at the end of construction or artificial ground freezing).

### 4.1.8 Pathways for groundwater flow

Groundwater control works can sometimes inadvertently create artificial pathways associated with wells, excavations or even a structure itself. Flow of groundwater along these pathways can affect groundwater quality, or the quantity of water available to groundwater sources.

Possible groundwater pathways include:

- **Vertical pathways via boreholes and wells:** wells and boreholes may puncture low permeability layers, increasing the risk of surface pollution finding its way down into the aquifer or allow mixing of shallow and deep groundwater in aquifers where groundwater quality is stratified. This is a particular issue if some strata may contain saline or contaminated water. Where flowing artesian conditions are present the construction of a well can allow water from the aquifer to flow freely at the surface. Specialist advice should be obtained when drilling in flowing artesian aquifers.
- **Vertical pathways via excavations and structures:** the excavations or structures being constructed also have the potential to create vertical pathways for groundwater flow. Where aquifer conditions are sensitive (for example close to significant water supply wells) deep structures such as shafts or basements should be designed to limit the potential for creation of vertical flow paths. This can be done by using raft foundations in preference to piles that would puncture low permeability aquitard layers. If piling or ground improvement methods have to be used, methods should minimise the formation of vertical flow paths. Guidance is given in Westcott *et al* (2001).

- **Horizontal pathways:** linear permeable features may also act as groundwater pathways. Examples include horizontal wellpoints (Section 2.1.4) or the granular bedding around sewer pipelines.

If the potential impacts from groundwater pathways are of concern a number of mitigation measures could be considered:

1. Design of wells and piezometers to include grout seals at appropriate levels to prevent vertical seepage around the outside of the well casing.
2. Well and piezometer casing to stand sufficiently above ground level to prevent surface waters being able to pass into the well casing (and from there into the aquifer) in the event of localised flooding around the well.
3. Top of the well casing to be capped when not in use, and sealed around the pumping equipment when in use. This will reduce the risk of pollution by substances flowing down the well.
4. Wells and piezometers to be appropriately sealed on completion (SEPA, 2010 and Environment Agency, 2012).

### 4.1.9 Impact of abstraction on water supply wells or springs

**See also**
5.2 Site investigation

In many parts of the UK groundwater is obtained from wells and springs as part of public water supplies and for private supplies for domestic and industrial users. If groundwater control is carried out near an existing well or spring abstractions, there is a risk that the abstractions will be 'derogated' – water levels may be lowered in wells, flow from springs may reduce and water quality may be adversely affected. The effect is often temporary, and may cease soon after the end of groundwater control pumping, but can cause considerable inconvenience and cost to groundwater users.

A desk study (see Section 5.2) is one possible way to identify any vulnerable water abstractions (well or spring sources), which may be impacted by a proposed groundwater control scheme. Environmental regulators (see Section 4.2) often hold records of the location of water sources through licensing schemes. It should be noted that under the various water abstraction regulatory schemes in the UK, not all water sources are licensed (many lower volume abstractions being below the lower volumetric limit of licensing schemes). So, the absence of licensed abstractions does not automatically mean the absence of groundwater sources operated by local users and communities. Larger sources (for example, public water supplies) should have Source Protection Zones (SPZs) delineated around them by the environmental regulators, within which engineering works require special permission. Information on SPZs can be obtained from the environmental regulators, either from published information on their websites, or by direct consultation with their staff. Where impacts on groundwater sources are of potential concern, numerical modelling or analytical calculations may be necessary to assess the potential impacts. Specialist hydrogeological advice should be obtained.

Depending on the scale of the predicted impact, possible mitigation measures include:

1. Survey of existing condition and performance of wells or springs, and implementation of a monitoring programme.
2. Replacement of the reduction in output of the water source with a temporary tanker supply.
3. Deepening of existing supply wells or lowering the pump level in the wells.
4. Drilling new water supply wells, either deeper, or in a different aquifer, or location, which is likely to be less impacted.
5. Construction of new water mains into the area, to provide a mains supply to replace the wells or springs.

## 4.1.10 Impact on groundwater dependent features

See also
2.3 Recharge systems
5.2 Site investigation

Groundwater plays an important role in supporting many surface water features. Many rivers receive a contribution to their flow from groundwater. Also wetlands and ponds (which can be important ecological habitats) are often the result of groundwater flows or springs. Collectively these are known as 'groundwater dependent features'.

Groundwater control schemes can impact on groundwater dependent features in one of two principal ways:

1. Pumping may draw water from the feature, for example by lowering groundwater levels so that water losses increase from the base of a pond or wetland.
2. Pumping (or groundwater barriers) may intercept (or block) water that would otherwise have reached the feature. For example, where groundwater pumping reduces baseflow reaching a river, thereby reducing flow in the river.

When assessing the potential impact on groundwater dependent features, a desk study (Section 5.2) can be useful in identifying nearby features. Many ecologically sensitive sites have been identified as being 'designated sites' by the environmental regulators and nature conservation bodies. Examples include Sites of Special Scientific Interest (SSSI) or sites designated within the Habitats Directive, such as Special Areas of Conservation (SAC) or Special Protection Areas (SPA). Such sites are protected by legislation and should not be damaged by the works or subject to adverse impact on the integrity of the site. If any such sites are identified in the vicinity of the works, specialist ecological and hydrological advice should be obtained.

It may be possible to reduce impacts by constructing a physical cut-off barrier between the dewatering system and the feature to be protected (Table 1.2). An alternative approach would be to use artificial recharge (see Section 2.3) of groundwater or surface water or to pipe a portion of the discharge water directly to the feature. The temperature, chemistry and sediment content of the discharge water (Section 4.1.3) must be assessed to ensure that there will not be adverse reactions with the water and ecology of the feature. Also, the risk of additional erosion must be considered (Section 4.1.2).

## 4.1.11 Saline intrusion

For large-scale infrastructure projects, the potential effect on water resources from temporary groundwater control measures may have to be assessed together with the longer-term impact of the permanent structures. The hydrogeological investigation may need to consider the intrusion of saline water into an aquifer. The techniques and models available to assess the impact on local and regional water resources and the potential risks of saline intrusion (Box 4.3) are beyond the scope of this guide, but further information can be found in Chapters 9 and 12 of Fetter (2014) or Chapter 9 of Bear (1979).

***Box 4.3 Case study of groundwater control to restrict saline intrusion***

A major road tunnel beneath a river in the south of England was constructed using the immersed tube technique, with concrete sections cast in a basin adjacent to the river channel (Leiper *et al*, 2000). The construction method gave rise to two possible sources of saline intrusion into the groundwater resource:

- the cutting of the trench across the river channel, into which the tunnel units were to be laid, removed the river bed silt exposing the underlying chalk aquifer to estuarine waters
- the water table in the casting basin was lowered for lengthy periods before being flooded with estuarine water to float the concrete tunnel sections out into the river; flooding the basin risked further saline intrusion into the chalk aquifer.

As part of the Act of Parliament which enabled the project, the regulatory authority obtained provisions for temporary works to protect the chalk aquifer. As a result the regulatory authority was fully consulted about the temporary works, including the proposed pumping flow rates. It became clear as the scheme progressed that the pumping flow rates originally envisaged would not be adequate, and the pumping rate was raised from approximately 300 l/s to 400 l/s. The distance of influence extended up to 5 km southwards from the casting basin.

Comprehensive monitoring of the groundwater control system included flow rates and groundwater levels, together with chemical testing of groundwater samples to assess the extent and movement of saline water in the aquifer. This enabled a detailed hydrogeological model of the area to be developed, which was used to design a pumping regime during the flooding of the basin to restrict saline intrusion.

## 4.1.12 Artificial recharge of groundwater

See also
2.3 Recharge systems
6.6 Settlement

At some sites it can be unacceptable to lower groundwater levels. The congested nature of our cities can mean that lowering of the groundwater level would lead to excessive settlement of sensitive structures or services, particularly of historic buildings (see Sections 4.1.6 and 6.6). Abstraction for construction works could also lead to depletion of aquifer levels, putting other groundwater sources at risk. As a result the requirement to maintain groundwater levels (typically by artificial recharge, see Section 2.3, or otherwise) could be specified as a mitigation measure by the environmental regulators during the abstraction licensing process. The design and operation of artificial recharge schemes is not straightforward, and specialist advice should be obtained. Specific environmental permissions may be needed to allow water to be recharged into the ground (see Table 4.3 and Section 4.2.3). An example of groundwater recharge being used to prevent aquifer depletion is given in Box 4.4.

*Box 4.4 Case study of groundwater recharge to prevent depletion of regional groundwater resource*

It was proposed to construct a tunnel beneath a river in East Yorkshire. The deepest part of the tunnel was to be 14 m below ground level and the temporary works were likely to involve the use of deepwells to lower piezometric levels in the underlying chalk aquifer. A public supply borehole was approximately 1.5 km from the site and calculation of the distance of influence indicated that the supply borehole might be significantly affected by the groundwater control system. After consultations with the project client and designer at the planning stage, the regulatory body prepared a draft consent, which set a discharge limit of 23 l/s for disposal to the river in order to reduce the effect on the supply borehole – all abstracted water in excess of that figure was to be recharged back to the chalk aquifer. The consent stated that if background groundwater levels fell below a specified drought control line, all abstracted water had to be recharged back to the aquifer until levels rose above the drought line. Because the regulatory body was involved at the planning stage, all contractors tendering for the works were aware of the special requirements. When construction started, recharge wells were installed with sufficient capacity to accept 100 per cent of the discharge flow rate in case drought conditions occurred during the construction period.

# 4.2 REGULATORY FRAMEWORK FOR GROUNDWATER CONTROL WORKS

In the UK the abstraction of groundwater is regulated by law; in general, any significant abstractions will require a licence or permit. The licence will set limits on the quantity of water, which can be abstracted, and may specify requirements to monitor abstracted volumes or to monitor groundwater levels. The licensing system and the associated monitoring allow the regulatory authorities to keep records of total abstractions from particular aquifers.

The discharge of groundwater from dewatering systems will also require some form of permit or licence. The principles of UK regulation of abstraction and discharge of groundwater are briefly summarised in Table 4.3 and outlined in the following sections.

Historically, during the latter part of the 20th century, groundwater abstraction for the purposes of dewatering (for construction, mining or quarry activities) and also land drainage was exempt from abstraction licensing. Although the regulators did not have direct powers to set limits on abstraction rates, or to require specific environmental mitigation measures, often on major or sensitive schemes they worked with the contractor or project team to incorporate reasonable measures into the design and implementation of groundwater control schemes that would limit the environmental impact. In the 15 years following adoption by the European Union of the EU Water Framework Directive (Directive 2000/60/EC), there was a requirement to widen the scope of the abstraction licensing process and to allow regulatory bodies to better manage water resources on a more holistic basis. Therefore, more stringent licensing requirements were introduced for dewatering systems, and are now a requirement across the UK.

The UK Regulators governing groundwater abstraction are:

- England: Environment Agency (EA)
- Wales: Natural Resources Wales (NRW)
- Scotland: Scottish Environment Protection Agency (SEPA)
- Northern Ireland: Northern Ireland Environment Agency (NIEA).

Additionally, where groundwater dependent ecosystems may potentially be impacted by groundwater abstractions, then statutory nature conservation regulators may need to be consulted. These are:

- England: Natural England (some water quality issues are managed under the Environment Agency remit)
- Wales: Natural Resources Wales
- Scotland: Scottish Natural Heritage (groundwater ecosystem protection administered by SEPA)
- Northern Ireland: Northern Ireland Environment Agency (Natural Heritage and/or water management sections).

Information on the policies and regulation of each regulator can be found on their websites, addresses for which are given at the end of this chapter.

A common problem on many groundwater control projects is that the parties involved do not allow sufficient time to apply for the necessary abstraction and discharge permissions and to carry out any associated studies and impact assessments (such as a hydrogeological impact appraisal, see Section 4.2.1). It should also be recognised that typically the application process for groundwater abstraction and water discharge are separate and independent processes. Note that where discharge is to a sewer then permission will be required from the relevant sewerage utility who themselves are subject to the same regulatory regime.

In addition to groundwater pumping for dewatering of excavations, pumping tests may be carried out as part of site investigation or dewatering trials (see Section 5.3.1). In some cases, especially for tests which involve large pumping rates, long pumping durations or which are in environmentally sensitive locations, abstraction and discharge permissions may be required. When planning pumping tests the environmental regulator for the site location should be consulted to determine any necessary permissions.

For complex and environmentally sensitive sites, the complete regulatory process can take up to several months to obtain the necessary permissions. For many construction projects it is unlikely that the contractor will have sufficient time to apply for a permission from scratch once the site works are about to begin. Ideally, initial discussions (sometimes termed 'early consultation') should be commenced, and applications should be lodged with the regulatory authorities by the project client or client's representatives at the planning stage. This allows the regulators additional time and any draft consent information can be incorporated into the tender documents so that potential contractors are aware of any likely constraints. The successful contractor should then finalise the consent with the regulators.

Note that application fees and other charges, including volumetric charges applied to the cumulative pumped volumes, may apply for both abstraction, discharge and other forms of consent. The fee structure is generally set by the relevant authority or regulatory body and may be subject to annual update. In some instances, particularly for discharge consents, costs can be significant.

*Table 4.3 Regulatory permissions for dewatering abstraction and discharge*

| Country | Regulator | Abstraction permission | Discharge permission |
|---|---|---|---|
| England | Environment Agency (EA) | • Governing regulations: Water Resources Act 1991 (as amended)<br>• Flow rate <20 $m^3$/day: no permission required. No requirement to notify EA<br>• Abstraction for 27 days or less, any flow rate > 20 $m^3$/day (water not put to use): temporary licence required<br>• Abstraction for 28 days or more, flow rate >20 $m^3$/day (water not put to use): transfer licence required<br>• Abstraction for 28 days or more, flow rate >20 $m^3$/day (water put to use): full licence required for portion put to use, transfer licence required for portion not put to use.<br>Note: the flow rate quoted for the different levels of licensing is the total dewatering flow rate from the site or project, not the flow rate from individual sumps, pumps, or wells. | ***Discharge to surface water or groundwater***<br>(EA) regulated under Water Resources Act 1991 (as amended).<br>Where dewatering pumping is for less than three months, the EA has granted an exception that dewatering flows can be discharged to surface waters without the need for an Environmental Permit, provided that the water is uncontaminated and will not cause adverse effects on aquatic life. In all other circumstances an Environmental Permit is required for discharge of dewatering flows to surface water or to groundwater.<br>If water treatment technologies are used to remove contaminants or otherwise treat water before discharge, a mobile plant permit and deployment form will also be required from the EA for the groundwater treatment plant under the Environmental Permitting (England and Wales) Regulations 2010.<br>***Discharge to sewer***<br>Trade Effluent Consent required (from Regional Water Company).<br>If water treatment technologies are used to remove contaminants or otherwise treat water before discharge, a mobile plant permit and deployment form will also be required from the EA for the groundwater treatment plant under the Environmental Permitting (England and Wales) Regulations 2010. |

continued...

...continued from

| Country | Regulator | Abstraction permission | Discharge permission |
|---|---|---|---|
| Wales | Natural Resources Wales (NRW) | • Governing regulations: Water Resources Act 1991 (as amended)<br>• Flow rate <20 $m^3$/day: no permission required. No requirement to notify NRW<br>• Abstraction for 27 days or less, any flow rate > 20 $m^3$/day (water not put to use): temporary licence required<br>• Abstraction for 28 days or more, flow rate >20 $m^3$/day (water not put to use): transfer licence required<br>• Abstraction for 28 days or more, flow rate >20 $m^3$/day (water put to use): full licence required for portion put to use, transfer licence required for portion not put to use.<br>Note: the flow rate quoted for the different levels of licensing is the total dewatering flow rate from the site or project, not the flow rate from individual sumps, pumps, or wells. | ***Discharge to surface water or groundwater***<br>(NRW) regulated under Water Resources Act 1991 (as amended).<br>Where dewatering pumping is for less than three months, the NRW has granted an exception that dewatering flows can be discharged to surface waters without the need for an environmental permit, provided that the water is uncontaminated and will not cause adverse effects on aquatic life. In all other circumstances an environmental permit is required for discharge of dewatering flows to surface water or to groundwater.<br>If water treatment technologies are used to remove contaminants or otherwise treat water before discharge, a mobile plant permit and deployment form will also be required from the NRW for the groundwater treatment plant under the Environmental Permitting (England and Wales) Regulations 2010.<br>***Discharge to sewer***<br>Trade effluent consent required (from regional water company).<br>If water treatment technologies are used to remove contaminants or otherwise treat water before discharge, a mobile plant permit and deployment form will also be required from the NRW for the groundwater treatment plant under the Environmental Permitting (England and Wales) Regulations 2010. |

continued...

...continued from

| Country | Regulator | Abstraction permission | Discharge permission |
|---|---|---|---|
| Scotland | Scottish Environment Protection Agency (SEPA) | Governing regulations: The Water Environment (Controlled Activities) (Scotland) Regulations 2011 (as amended)<br>Dewatering abstractions that comply with General Binding Rule 15 (GBR15) require no notification to SEPA provided that GBR15 is complied with. The requirements of GBR15 can be summarised as:<br>• in geological strata where groundwater flow rates are low (eg silts, clays, low permeability bedrock), groundwater may only be abstracted for less than 180 consecutive days<br>• in geological strata where groundwater flow is high (eg sands and gravels and sandstones) groundwater may only be abstracted for a total of five separate days, in any 180 consecutive day period<br>• groundwater must not be abstracted within 250 m of a wetland or with 250 m of any (non-dewatering) abstraction<br>• all reasonable steps must be taken to ensure that the quantity of sediment in the abstracted water is minimal<br>• discharge of the abstracted water must be via a surface water drainage system authorised by SEPA or Scottish Water (as appropriate).<br>Dewatering abstractions that do not comply with GBR15 are subject to the following requirements:<br>• flow rate <10 $m^3$/day: compliance with General Binding Rule 2 (GBR2) is required. No requirement to notify SEPA provided that GBR2 is complied with<br>• flow rate 10 to 50 $m^3$/day: registration with SEPA required<br>• flow rate 50 to 2000 $m^3$/day: simple licence required<br>• flow rate >2000 $m^3$/day, or boreholes greater than 200 m depth: complex licence required.<br>Note: the flow rate quoted for the different levels of licensing is the total dewatering flow rate from the site or project, not the flow rate from individual sumps, pumps, or wells. | ***Discharge to surface water or groundwater***<br>(SEPA) regulated under The Water Environment (Controlled Activities) (Scotland) Regulations 2011 (as amended).<br>No requirement to notify SEPA for discharge of uncontaminated groundwater abstracted directly from a dewatering system without contact with any other drainage run-off. This only applies to uncontaminated groundwater.<br>For contaminated discharges:<br>• discharge rate < 10 $m^3$/day: registration with SEPA required<br>• discharge rate 10 to 100 $m^3$/day: simple licence required<br>• discharge rate >100 $m^3$/day: complex licence required.<br>***Discharge to sewer***<br>Trade Effluent Consent required (from Scottish Water). |

continued...

...continued from

| Country | Regulator | Abstraction permission | Discharge permission |
|---|---|---|---|
| Northern Ireland | Northern Ireland Environment Agency (NIEA) | Governing regulations: Water Abstraction and Impoundment (Licensing) Regulations (Northern Ireland) 2006<br>• flow rate <10 $m^3$/day: compliance with permitted controlled activities (PCA) is required. No requirement to notify NIEA provided that PCAs are complied with<br>• flow rate 10 to 20 $m^3$/day: compliance with PCAs is required. NIEA must be notified<br>• flow rate 20 to 100$m^3$/day: simple licence required<br>• flow rate >100 $m^3$/day: complex licence required.<br>Note: the flow rate quoted for the different levels of licensing is the total dewatering flow rate from the site or project, not the flow rate from individual sumps, pumps, or wells. | ***Discharge to surface water or groundwater***<br>Trade effluent and site drainage discharge consent required under the Water (Northern Ireland) Order 1999 (NIEA)<br>***Discharge to sewer***<br>Trade Effluent Consent (from Northern Ireland Water) |

**Note**

Required permissions correct at January 2014. The regulators should be consulted to confirm current requirements at the time of the project, including the current GBRs and PCAs. Note that application fees and other charges (including volumetric charges applied to the cumulative pumped volumes) may be applicable and can prove significant in some instances.

## 4.2.1 Hydrogeological impact appraisal

On many projects where there is potential for adverse environmental impacts from dewatering activities, it may be necessary to formally assess the impacts. This process is known as hydrogeological impact appraisal or HIA. The requirement for a HIA may be specified by the regulator as a requirement for being granted a licence or permit, or may have been specified by the project's technical advisors to ensure groundwater impacts are adequately assessed prior to dewatering. The Environment Agency's standard methodology for HIA is given in Boak *et al* (2007).

## 4.2.2 Abstraction of groundwater

All groundwater control systems, which pump water, will require some form of permission from the relevant environmental regulator. Typically, the types of permission are structured in a 'tiered' manner, whereby 'significant' abstractions require more complex permissions than more 'minor' abstractions, where lesser permissions are needed – which might be as simple as registration with the regulator.

In general, regulators are likely to assess dewatering abstractions as significant and requiring more complex levels of permission if some of the following conditions exist:

- the proposed period of dewatering pumping is long (eg longer than one month)
- the proposed pumped flow rate is relatively large
- there are existing water users (eg water wells or spring supplies) in the area of the site, which may be affected by the dewatering pumping (Section 4.1.9)
- there are groundwater dependent features in the area of the site which may be affected by the dewatering pumping (Section 4.1.10)
- the regulator has concerns that groundwater resources in the area are at risk, either in terms of quantity (eg availability of groundwater resources) or quality (eg saline intrusion or migration of contaminated groundwater).

The necessary permissions for abstraction of groundwater for dewatering purposes in each part of the UK are summarised in Table 4.3. Typically, the more complex permissions will set limits on the quantity and rates of water that can be abstracted, and will specify monitoring requirements.

It should be noted that where the levels of abstraction licensing required by the regulatory authorities are tiered based on flow rate, the value to be used is the total dewatering flow rate from the site or project, not the flow rate from individual sumps, pumps, or wells.

## 4.2.3 Discharge of groundwater

**See also**
2.3 Recharge systems

There are four principal means of disposal of pumped water from dewatering systems (volumes generated are normally too great to be stored on site):

1 **Discharge to surface water**

In general, the disposal of pumped water from groundwater control systems to surface water including rivers, lakes and the sea or to groundwater (collectively referred to as 'controlled waters' under the regulatory regime in England and Wales) will require permission from the environmental regulator. The necessary permissions for discharge to surface water in each part of the UK are summarised in Table 4.3. The permissions may set limits on the quantity and quality of water that can be discharged, and may specify monitoring requirements.

Where water is to be discharged into a navigable waterway an Outfall Consent may be required from the relevant authority (eg the relevant port authority or the Canal & Rivers Trust or Scottish Canals for inland water ways and canals). The relevant waterway authorities are primarily concerned with potential impact on navigation, river banks and water levels (for canals and docks). Issues of water quality and flood defence generally remain within the remit of the environmental regulator.

2 **Direct discharge to groundwater (artificial recharge)**

Because the pumped water from dewatering systems is comprised of groundwater, it is sometimes assumed that the discharge of dewatering flows into the ground (eg by artificial recharge or deep borehole soakaway) does not require permission from the environmental regulators. This is categorically not the case, and in almost all cases permissions will be needed. The necessary permissions for discharge to groundwater in each part of the UK are summarised in Table 4.3. The permissions may set limits on the quantity and quality of water that can be discharged, and may specify monitoring requirements.

3 **Discharge to sewer**

In some areas it may be possible to dispose of water into the sewerage network. If it is to be disposed of to a sewer, the abstracted water from a groundwater control system is legally classified as trade effluent. Permission must be obtained from the sewerage utility before this can be done. The relevant sewerage utilities are:

- England and Wales: regional water companies
- Scotland: Scottish Water
- Northern Ireland: Northern Ireland Water.

Permissions normally take the form of a trade effluent consent or a trade effluent agreement, which may set limits on the quantity and quality of water that can be discharged, and may specify monitoring requirements. Charges (based on a cost per cubic metre) are generally levied for disposal of water in this way. For long duration discharge of substantial volume these charges can add up to a considerable sum.

4 **Discharge to land**

Occasionally, the discharge from a groundwater control system may be disposed of to land (for example by irrigation or spreading over fields, or to shallow soakaway). Environmental regulators may deem this a groundwater activity or a waste activity and the relevant environmental permission may be required.

## 4.3 KEY REFERENCES

BOAK, R, BELLIS, L, LOW, R, MITCHELL, R, HAYES, P, McKELVEY, P and NEALE, S (2007) *Hydrogeological impact appraisal for dewatering abstractions*, Science Report SC040020/SR1, Environment Agency, Bristol (ISBN: 978-1-84432-673-0)

PREENE, M and BRASSINGTON, F C (2003) "Potential groundwater impacts from civil engineering works" *Water and Environmental Management Journal*, vol 17, 1, Wiley, London, pp 59–64

## 4.4 REGULATOR WEBSITES

### *Abstraction and discharge of groundwater*

Environment Agency (EA): **https://www.gov.uk/government/organisations/environment-agency**

Natural Resources Wales (NRW): **https://naturalresources.wales**

Scottish Environment Protection Agency (SEPA): **www.sepa.org.uk**

Northern Ireland Environment Agency (NIEA): **www.doeni.gov.uk/niea**

### *Statutory nature conservation bodies*

Natural England: **https://www.gov.uk/government/organisations/natural-england**

Natural Resources Wales: **https://naturalresources.wales**

Scottish Natural Heritage: **www.snh.gov.uk**

Northern Ireland Environment Agency: **www.doeni.gov.uk/niea**

# 5 Site investigation requirements

## 5.1 OBJECTIVES OF SITE INVESTIGATION

**See also**
3.2 CDM Regulations
Chapter 6 Design
Chapter 7 Case studies

Site investigation is essential in order to describe the ground and groundwater conditions adequately; to identify potential groundwater problems; and, ultimately, to allow groundwater control measures to be designed.

Unfortunately, many investigations are designed mainly to provide the information required by the permanent works designer and not that required for the design and implementation of groundwater control for temporary works. Any groundwater control project carried out without an investigation directed, at least in part, to temporary works runs a risk of delays, additional costs or even redesign of the works.

Box 5.1 shows a case study where a poorly planned investigation, using boreholes of inadequate depth, did not identify a groundwater problem which caused a shaft to be abandoned during construction. The problems resulting from inappropriate investigations are highlighted in two reports by the ICE (1991) and from the Site Investigation Steering Group 1993) which concluded:

*Having a cheap or small scale site investigation does not necessarily save money.*

*You pay for a site investigation whether you have one or not.*

This chapter draws attention to the specific investigation requirements of projects where some form of groundwater control may be required. Objectives of investigation are summarised in Table 5.1, but a range of expertise may be needed for the project (Figure 5.1), and specialist advice should be taken. Chapter 6 outlines the technical aspects of groundwater modelling and the design of groundwater control systems and illustrates how the information from investigations is used.

Under the CDM Regulations (see Section 3.2) all parties involved in a construction project (including clients, designers and contractors) must work together from an early stage in order that safe methods of working are planned and adopted. Groundwater control can be vital for safe working below the natural groundwater level. To fulfil their duties under the CDM Regulations, clients and their designers must make sure that the organisations designing and carrying out the investigation are provided with details of the proposed excavation (eg depth, size, excavation, support methods). Even if full details have not been finalised, provisional information will help the investigation to be adequately directed. A flexible, phased investigation with regular communication between the client and those directing the investigation is the best approach.

Communication should be in both directions. Changes in the permanent or temporary works may require additional investigation, and the ground and groundwater conditions revealed can affect the feasibility and economics of the proposed works, possibly requiring redesign. Several of the case studies in Chapter 7 describe situations where working methods had to be changed at a late stage because of groundwater problems; some of these could have been avoided by better communication and direction at the investigation stage.

*Box 5.1 Case studies of inadequate site investigation for shaft construction (after Site Investigation Steering Group, 1993)*

A shaft 7.8 m deep was constructed through alluvial and glacial soils. Pre-construction investigation comprised two boreholes to 8.45 m and 8.65 m, but the site investigation contractor was not made aware of the details of the proposed construction works. The boreholes revealed clay containing silt and sand partings. No water strikes were recorded and no piezometers were installed. From the level of a nearby river and local topography, the investigation concluded that the standing groundwater level was probably within 2 m of ground level. The shaft was sunk using groundwater control by sump pumping from the shaft bottom. Base failure occurred with a massive ingress of water during excavation of the deepest shaft ring. The resulting damage led to the abandonment of the shaft.

### Ground conditions shown in boreholes

Post-failure boreholes revealed that a significant thickness of silty sand underlay the laminated clay about 1.6 m below shaft base level. Base failure occurred because of unrelieved water pressure in the sand; the lack of appropriate groundwater control measures stemmed from inadequate pre-construction investigation. For the estimated groundwater level and depth of excavation, an investigation depth of at least 15 m would have been appropriate, but the client did not seek the advice of a geotechnical specialist. On the basis of the post-failure investigation, a replacement structure was successfully constructed using an ejector well system to reduce water pressures in the silty sand and prevent base failure.

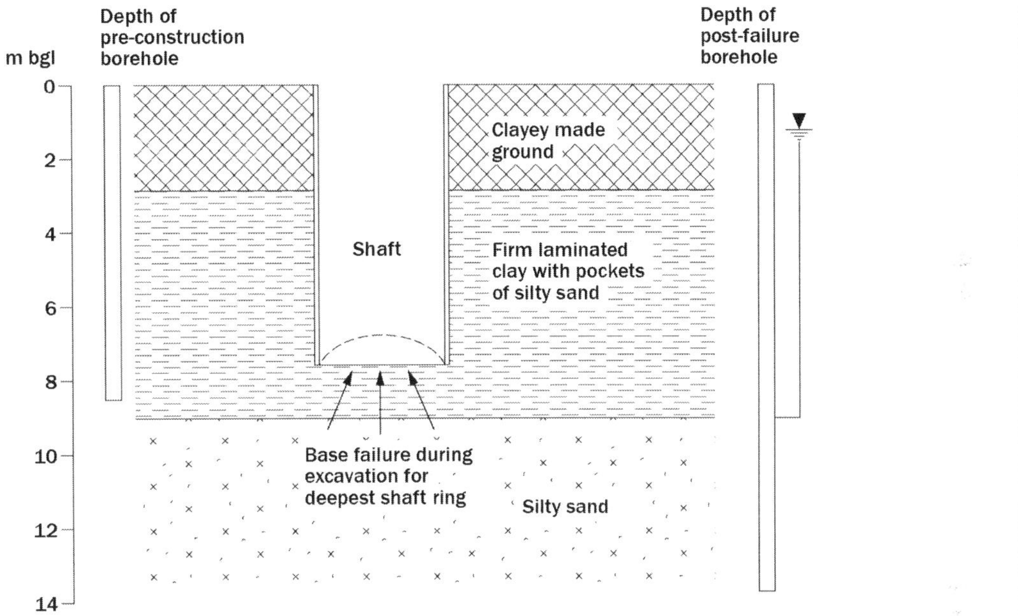


*Table 5.1 Site investigation objectives for a groundwater control project*

| Aspect | Objectives and information needs |
|---|---|
| Ground profile and stratigraphy | Carry out borehole surveys to 1.5 to 2 times the depth of the proposed excavation<br>Identify the presence and location of any water-bearing strata and low permeability layers |
| Hydrogeological parameters | Identify strata that may act as confined or unconfined aquifers, aquitards or aquicludes<br>Estimate permeability (and, if necessary, storage coefficients) for aquifers |
| Groundwater levels and pore water pressures | Determine by monitoring piezometers (long-term monitoring may be required) |
| Site and area conditions | Highlight possible aquifer recharge sources (eg rivers)<br>Identify any adverse environmental effects of groundwater control (eg settlement or effect on nearby water supply wells) |
| Geotechnical parameters | Assess coefficients of soil consolidation and compressibility to determine if settlement of nearby structures is a problem (especially if peat or soft clay are present) |
| Groundwater chemistry and contamination | Check for aggressive groundwater conditions (eg dissolved iron, hardness or chloride levels) or for possible contamination<br>Review site history and nearby land use |

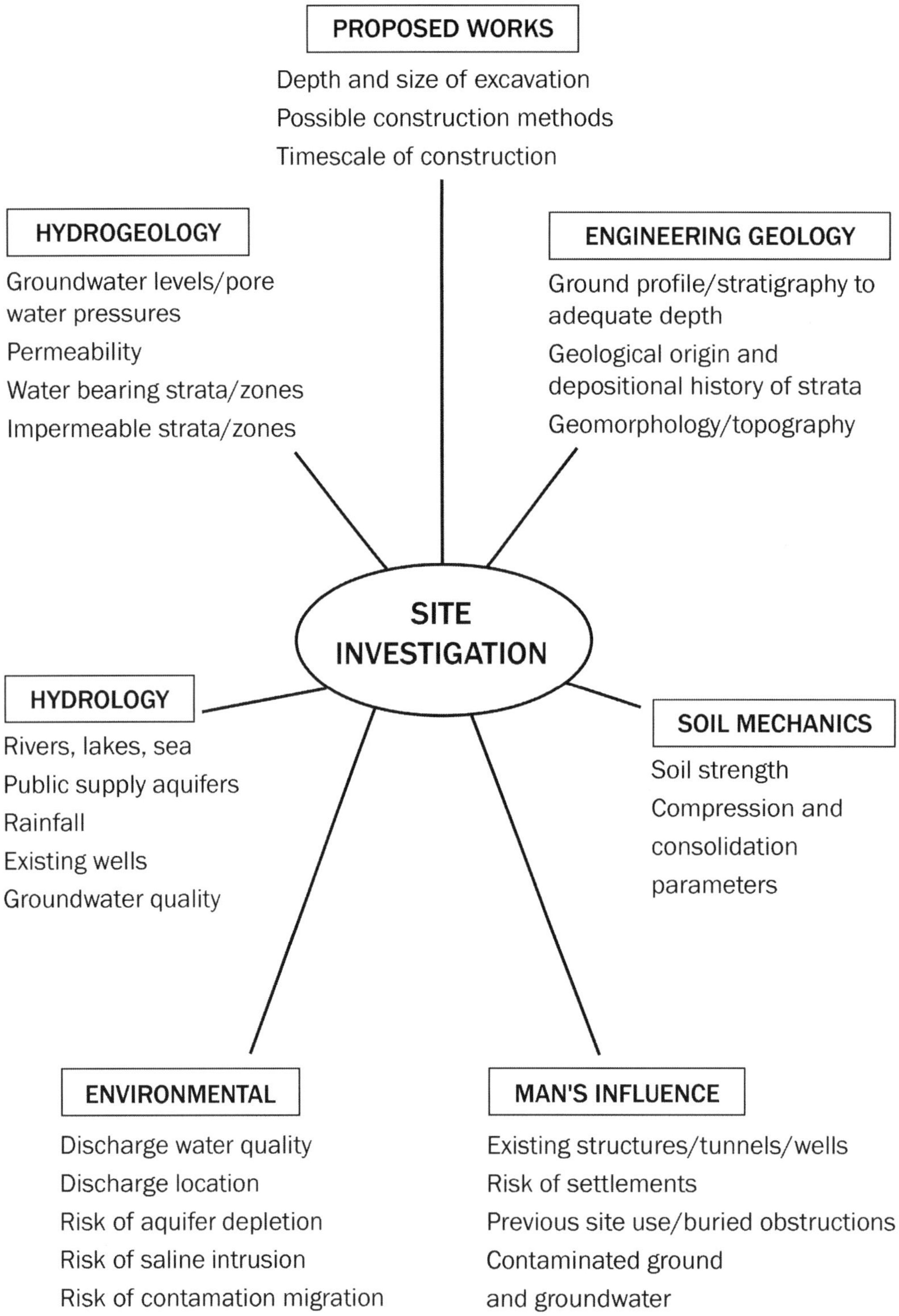


***Figure 5.1*** ***Information needs to be considered in site investigation for groundwater control projects***

All ground investigation work is specialised and should be carried out only by organisations with the necessary specialist expertise and equipment. The requirements for geotechnical specialists and advisors are given in Site Investigation Steering Group (1993, 2011, 2013), together with advice on selection of the geotechnical team and recommendations about the client's obligations. Further information on the planning and execution of site investigations is given in BS 5930:2015, BS EN 1997-2:2007, Clayton *et al* (1995) and Weltman and Head (1983). Methods of permeability testing are described in BS EN ISO 22282:2012. Safety guidance for investigation sites is given in BDA (2015). Recommendations for investigations on contaminated or potentially contaminated sites are given in Harris *et al* (1995).

## 5.2 SITE INVESTIGATION METHODS

The principal phases of a site investigation are:

1 Desk study.

2 Site reconnaissance.

3 Ground investigation.

4 Laboratory testing.

5 Reporting.

The desk study and site reconnaissance are sometimes overlooked in investigations, but are essential to gathering information about the project and investigating the background to the site. This stage can involve study of geological and hydrogeological maps and records, water supply well records, walkover inspection of the site and study of previous construction experience at the site (if available). Details of the proposed works will be required to allow the ground investigation to be designed. The importance of the desk study and site reconnaissance cannot be overstated; the ground investigation can only be adequately designed on the basis of the results of this stage. A failure to recognise potential groundwater problems here can be one of the main reasons for poor or inadequate investigations. It is important that the potential risk of the presence of groundwater contamination is assessed at this stage.

The ground investigation involves fieldwork on site. Laboratory testing may be carried out on samples recovered during ground investigation. Methods of ground investigation relevant to groundwater control projects are given in Table 5.2. The fieldwork may be carried out in several phases, with each phase designed using the information gathered previously. For example, if initial boreholes reveal the presence of a major water-bearing stratum, which may need dewatering, later investigation could take the form of pumping or permeability tests. Long-term monitoring of groundwater levels in piezometers or standpipes may need to continue beyond the initial fieldwork period.

The site investigation report describes the site, the work carried out and the results obtained. This information is normally given in a factual report, but, additionally, an interpretative report should be commissioned from a geotechnical specialist which will include engineering recommendations for the proposed project. If groundwater problems are identified, these should be discussed in the report. At investigation stage construction methods are unlikely to be finalised, but it is essential that potential groundwater problems are highlighted.

The identification of potentially water-bearing or low permeability strata and location of the water table or piezometric level in relation to the proposed excavation is the starting point for any assessment of groundwater control needs. This will generally involve the study of borehole and trial pit logs and groundwater level records from boreholes and piezometers.

Table 5.2 *Methods of ground investigation*

| Method | Parameters obtained | Notes |
|---|---|---|
| Boreholes (cable percussion or rotary drilling) | Ground and groundwater profile | • allows sampling of soils<br>• allows *in situ* permeability testing<br>• gives indication of groundwater levels (especially if a piezometer or standpipe is installed)<br>• may allow groundwater sampling |
| Trial pits | Ground and groundwater profile | • depth limited to 3 m to 5 m<br>• may be difficult to progress below groundwater level<br>• allows *in situ* stability of soil to be observed<br>• can allow representative sampling of very coarse soils<br>• gives indication of groundwater levels<br>• may allow groundwater sampling |
| Static cone penetration testing | Ground and groundwater profile | • penetration may be difficult in coarse granular soils, stiff clay or weak rock<br>• can give detail of soil profile<br>• piezocone can record pore water pressures and carry out permeability tests in fine-grained soils |
| Pumping tests and observation piezometers | Groundwater level, permeability and aquifer parameters | • often the most reliable method of obtaining permeability values<br>• allows groundwater samples to be obtained |
| Geophysics | Ground profile | • can provide information between widely spaced boreholes |

## 5.2.1 Ground profile

**See also**
4.1 Environmental impacts
Box 6.3 Aquifers
6.6 Settlement

Correct identification of the sequence of strata is often crucial for groundwater control projects. The investigation should describe the relationship between any aquifers, aquitards and aquicludes. The requirements for engineering description of soils and rocks for borehole and trial pit logs are given in BS EN 14688-1:2002, BS EN 14689-1:2003 and Norbury (2010).

The soil description, together with groundwater level information and any permeability tests (Section 5.3) should identify which strata are water-bearing (and may form aquifers) and which strata are of low permeability (aquicludes and aquitards). In combination with groundwater level records, the investigation should allow any potential aquifers to be classified as confined or unconfined (Box 6.3). The ground profile should also identify any compressible strata and potential settlement problems (Section 6.6) or any other adverse environmental effects of groundwater control (Section 4.1).

## 5.2.2 Groundwater levels

**See also**
Box 3.2 Piezometer monitoring

The investigation should determine as reliably as possible the groundwater level in relation to the level of the proposed excavation. However, there are problems with accurately determining groundwater levels. Water levels during boring can be affected by the drilling methods, so the most reliable way is by installation and monitoring of standpipes or piezometers (see Box 3.2). A standpipe (Figure 5.2) is an open tube, perforated over part of its length, intended to respond to groundwater levels over its full depth. In a standpipe piezometer (Figure 5.2) bentonite or grout seals are used to isolate the perforated section so that the device responds to a specific zone or stratum. In some circumstances, such as in very low permeability silts or clays, or where groundwater levels or pore water pressures may vary rapidly, specialist equipment such as electronic transducers linked to dataloggers may be installed. More recently grouted in place vibrating wire transducers (VWT) have proved to be an effective technique for monitoring water pressures at multiple levels in a single borehole (Yungwirth *et al*, 2013). Care must be taken during design and installation of standpipes and piezometers to select the correct instrument type and depth of response zones appropriate to the groundwater regime. Following installation, testing and commissioning may be required (eg purging or development to remove dirty water from standpipe tubing). Advantages and disadvantages of various methods of determining groundwater levels are summarised in Table 5.3.

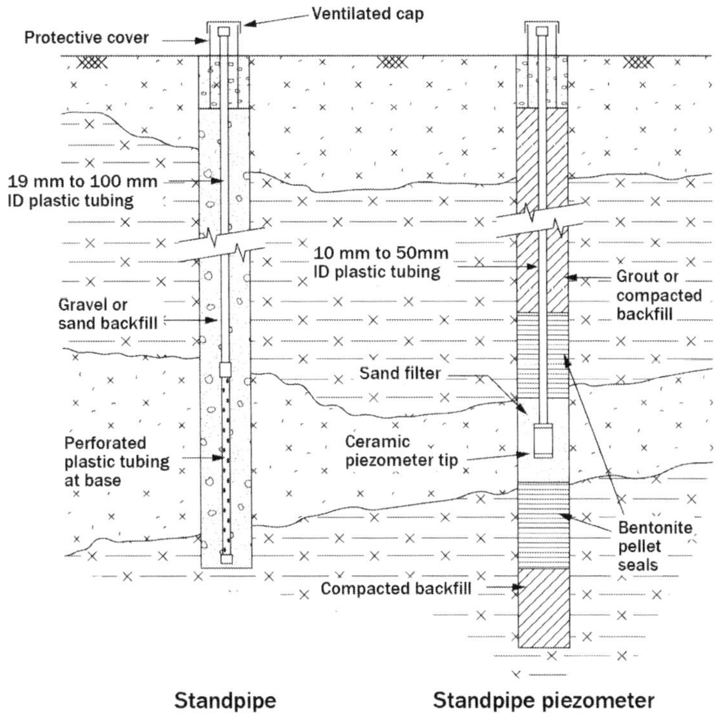


*Figure 5.2 Standpipe and standpipe piezometer (after Clayton et al, 1995)*

Groundwater levels on a site may vary with time as a result of various factors:

- tidal or river flood effects
- seasonal and climatic effects (including periods of drought or heavy rainfall)
- barometric effects
- pumping from nearby water supply wells
- pumping from nearby groundwater control systems.

The resulting changes in groundwater levels significantly influence assessment of the groundwater control requirements at a site. The site investigation desk study and site reconnaissance should highlight relevant factors, but the effects can only be assessed if long-term monitoring of standpipes or piezometers is carried out (perhaps by using piezometer datalogging equipment on unattended sites).

*Table 5.3 Methods of determining groundwater levels*

| Method | Notes |
|---|---|
| Observation piezometers (standpipe piezometers, pneumatic piezometers, hydraulic piezometers, electronic transducers) | • monitoring of piezometers is the best way of determining groundwater levels<br>• may need to be developed or purged before use<br>• a standpipe piezometer should periodically be pumped out or topped up with clean water to check the water returns to its true level. |
| Observation standpipes and wells | • not sealed into a specific stratum so may give confusing results if more than one water-bearing layer is present or if pore water pressures are not hydrostatic<br>• may need to be developed or purged before use<br>• a standpipe should periodically be pumped out or topped up with clean water to check the water returns to its true level. |
| Borehole records during drilling | • recorded water levels do not necessarily represent groundwater levels because of drilling rate, method and addition or removal of water by driller or the use of muds in rotary boreholes<br>• during rapid drilling in fine-grained soils water entries or strikes may be missed<br>• standing water levels may be incorrect unless adequate time is allowed for equalisation. |
| Trial pit records | • seepage into excavation can be observed directly<br>• standing water levels may be incorrect unless adequate time is allowed for equalisation. |

## 5.3 PERMEABILITY TESTING

*Assessment of soil permeability is important in any investigation for projects involving excavation below groundwater level. However, obtaining appropriate and representative values of permeability is difficult because the ground is likely to be anisotropic and heterogeneous, so permeability varies from one point to the next. Even if it could be obtained, there is no single value of permeability in the ground waiting to be measured.*

**See also**
Table 1.1 Permeability values
6.1.3 Permeability selection
Table 6.2 Reliability of permeability methods

The problem of determining permeability is compounded by limitations of the testing techniques used and their interpretation. A groundwater control system pumping from an aquifer may cause drawdowns over a wide area of influence (perhaps several hundred metres across), and so affects a large volume of soil. One of the most reliable ways of estimating the effective permeability of a large mass of soil is by a well pumping test: a well is pumped and the flow rate and the drawdown in surrounding observation wells are recorded; appropriate analysis can provide estimates of mass permeability. Some other permeability tests are based on only a very small volume of *in situ* soil in boreholes, or use samples recovered from boreholes and tested in the laboratory. There are risks and complications in scaling up from individual test results to the permeability values to be used in design.

It is important that permeability tests or sampling for testing are carried out at appropriate locations. Permeability should be estimated for those strata (principally potential aquifers) which will be affected by any groundwater control. Tests should be carried out below groundwater level; otherwise the results are useless. If the aquifer extends considerably below the planned depth of excavation, permeability tests should be made to the depth likely to be significantly affected by pumping (typically 1.5 to 2 times the depth of excavation).

Methods of estimating permeability are given in Table 5.4. More detailed notes and references on commonly used tests are given in the following sections. The problem of selecting values of permeability to be used in design is discussed in Section 6.1.3, and Table 6.2 gives guidance on the reliability of permeability estimates from various methods.

Permeability of naturally occurring soils covers a very wide range, from less than $10^{-9}$ m/s for intact clays to more than $10^{-1}$ m/s for open gravels and cobbles. It is not a simple matter to give guidance on the approximate permeability of soil types, because in addition to the soil description and grading, soil fabric (such as layering or fissuring) affects permeability. Permeability may be anisotropic, with the horizontal permeability, $k_h$ , often greater than the vertical permeability, $k_v$ . Table 1.1 gives very approximate ranges of soil permeability. This section and the following sections consider estimation of permeability for soils in the range where groundwater control measures are generally applied. Tests in very low permeability soils such as intact clays are not considered; special tests and analyses may be required in such circumstances (eg Brand and Premchitt, 1982).

**See also**

6.1.3 Permeability selection

Table 6.2 Reliability of permeability methods

*Table 5.4 Methods of estimating permeability*

| Category | Method | Notes |
|---|---|---|
| *In situ* (large scale) | Well pumping tests (Section 5.3.1) | • can estimate the permeability of a large volume of soil<br>• can provide information on boundary conditions. |
| | Groundwater control trials | • generally only cost-effective for large projects. |
| *In situ* (small scale) | Borehole tests (Section 5.3.2), eg falling head test, rising head test, constant head test, packer test (Section 5.3.4) | • test only a small zone around borehole<br>• can be dramatically affected by soil disturbance due to drilling – disturbance effects may lead to underestimates of permeability<br>• packer testing normally only carried out in rock. |
| | Piezometer tests (Section 5.3.3), eg falling head test, rising head test, constant head test | • test only a small zone around piezometer. |
| | Specialist tests, eg piezocone, *in situ* permeameter | • generally used only in fine-grained soils (silts or clays). |
| Inverse numerical modelling | | • Uses groundwater monitoring data, eg piezometer readings, to back analyse permeability |
| Laboratory | Particle size analysis (Section 5.3.5) | • permeability interpreted from grading curves<br>• results are dependent on quality of samples obtained<br>• loss of fines or mixing of layered soils can affect results dramatically<br>• not representative for structured or very fine grained soils. |
| | Permeameter testing, eg triaxial cell, Rowe consolidation cell, oedometer consolidation cell | • results likely to be affected by sample disturbance<br>• soil fabric and structure may mean sample size affects results. |
| Virtual assessment | | • Can give approximate guide to permeability to be used to corroborate results from other tests |

## 5.3.1 Well pumping tests

**See also**

6.1.4 Numerical modelling

A well pumping test is conducted by pumping from a well and measuring the discharge flow rate and the drawdown of groundwater levels in an array of piezometers or observation wells radiating out from the well (Box 5.2). Pumping tests are more complex and expensive to carry out than borehole permeability tests but because they test a much larger volume of soil, they can give much more reliable estimates of permeability. There is no such thing as a standard pumping test, and sometimes relatively simple tests (of short duration and with comparatively few observation wells) can provide useful information very economically. Guidance for the planning and execution of tests is given in BS ISO 14686:2003 and BS EN ISO 22282-4:2012. Appropriate methods of analysis should be used to calculate hydrogeological parameters (eg permeability, storage coefficients and aquifer boundary conditions); guidance is given in Kruseman and De Ridder (1990) and Preene and Roberts (1994). The results of pumping tests can provide valuable information for the setting up or calibration of numerical groundwater models (see Section 6.1.4).

Geophysical methods are sometimes used to determine characteristics of the test well; available methods are described in BS 7022:1988. Flow logging may be particularly useful for identifying flow from fissured zones in rock.

Sometimes several wells, wellpoints or ejectors may be test pumped simultaneously in the form of a dewatering trial. Such trials are more expensive than pumping tests, but may be appropriate for large projects where groundwater control is critical. An example of a dewatering trial is documented by Powrie and Roberts (1990). Where artificial recharge (see Section 2.3) is planned, it may be appropriate to carry out a recharge pumping test, where water is re-injected into a test well for an extended period, with flow rate and groundwater levels monitored in the same way as for a conventional pumping test, see Roberts and Holmes (2011).

*Box 5.2 Well pumping test*

**Well pumping test**

A comprehensive well pumping test may consist of the following phases. Not all phases will be needed for all tests and recovery periods will be required between each phase:

1 Pre-pumping monitoring.
2 Equipment test.
3 Step-drawdown test (typically lasts 4 to 8 hours).
4 Constant rate pumping phase (typically lasts 1 to 7 days).
5 Recovery phase (typically lasts 1 to 3 days).

In general, the following parameters must be monitored:

- groundwater levels in pumped and observation wells (by datalogger or frequent manual dipping)
- discharge flow rate (by weir tank or flowmeter)
- groundwater quality (samples to be taken for analysis).

Test methods are described in BS EN 14686:2003 and BS EN ISO 22282-4:2012.

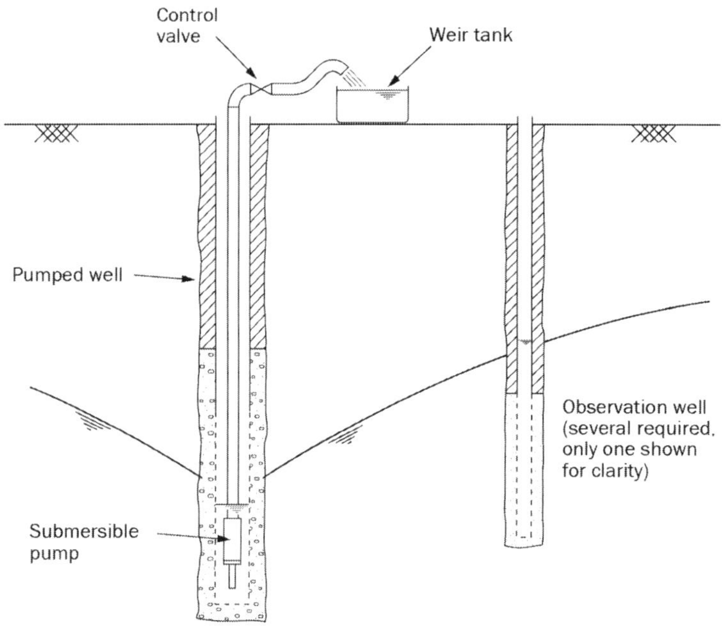


The following factors should be taken into account when designing well pumping tests:

1 Groundwater levels should be monitored prior to the test. If significant tidal or other variations are apparent, continuous monitoring over several days should be carried out.

2 The drawdowns measured in piezometers within 10 m to 20 m of the test well should ideally be at least 10 per cent of the required drawdown for the proposed groundwater control works. If the site is subject to tidal or background variations, the drawdown achieved should be significant relative to these fluctuations. However, if possible the test should model the flow conditions likely to occur during groundwater control. If the full-scale system is intended to reduce the piezometric head in a confined aquifer so that it becomes unconfined, the pumping test should be designed to create locally unconfined conditions if possible.

3 Confined aquifers respond to pumping much more rapidly than unconfined aquifers and may need very frequent monitoring during the early part of the test.

4 Ideally, pumping should be continued at least until steady-state drawdown conditions are achieved near the well.

5 A sufficient number of observation wells or piezometers are required to fully identify drawdown patterns around the well. If ground conditions are likely to vary around the well, or if there is a significant groundwater flow across the site, lines of observation wells should be located on more than one side of the well.

## 5.3.2 Falling, rising and constant head tests in boreholes

These *in situ* tests are carried out in boreholes (during or soon after drilling) and measure the permeability locally around the bottom of the borehole by inducing flow into or out of the ground (Box 5.3). The tests are described in detail in BS EN ISO 22282-2:2012. Once the initial water level in the borehole has been recorded, water is either added or removed and the rate at which the water in the borehole recovers to its original level (falling and rising head tests) is measured; in constant head tests the induced head and flow rate are measured. Unfortunately, clogging or silting up of the borehole often occurs in these tests, leading to inaccurate results. It is good practice to carry out both rising and falling head (or inflow and outflow) tests in the same borehole and to compare the results. In any event, results from these tests should be used with caution until corroborated by permeability estimates from other methods.

The following factors affect permeability estimates from tests in boreholes:

1 Results from borehole tests are not as reliable as from pumping tests because only a small zone of soil is tested.

2 Soil disturbance (such as particle loosening, compaction or smearing of silt and clay layers) caused by boring can affect results. It is important to try to clean out the bottom of the borehole prior to the test, but this can be difficult in practice.

3 Falling head tests are very prone to clogging or silting up at the bottom of the borehole when water is added. Permeability can be underestimated by several orders of magnitude. Results of falling head tests should be viewed with extreme caution unless corroborated by other, more reliable methods.

4 During a rising head test flow into the borehole may cause piping or boiling at the base, leading to overestimates of permeability. Rising head tests can also be prone to silting up if sediment is allowed to settle in the borehole. Constant head inflow tests suffer from many of the same drawbacks as falling head tests. Permeability estimates should be treated with caution.

5 For accurate test analysis, sufficient monitoring should be carried out to determine initial groundwater levels. Analysis is difficult if the groundwater level varies during the test (as a result of tidal effects, for example). Falling head tests where the groundwater level is close to ground level may need the borehole casing to be extended above ground level to create sufficient initial head.

6 Tests may not be possible in very permeable soils (greater than about $10^{-3}$ m/s) because water cannot be added or removed quickly enough to change the level in borehole.

*Box 5.3* ***Falling and rising head tests in boreholes***

**Falling head test**

Water added to raise level in borehole and induce flow from the borehole into the ground

*H* = excess head at time *t*

$H_o$ = excess head at *t* = 0

*A* = cross sectional area of borehole casing (excess head measured relative to initial groundwater level)

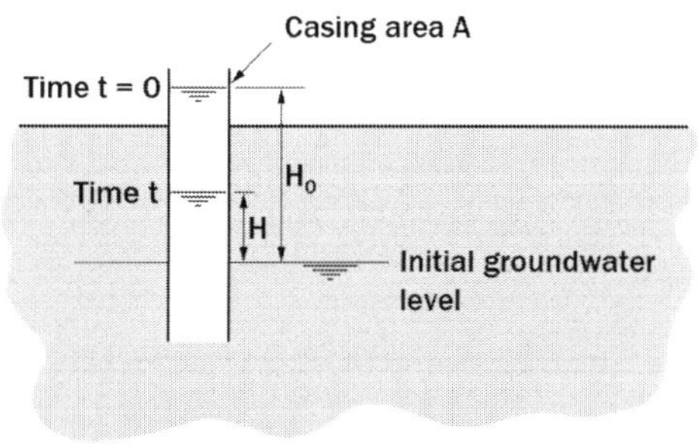


**Rising head test**

Water removed to lower level in borehole and induce flow from the ground into the borehole

*H* = excess head at time *t*

$H_o$ = excess head at *t* = 0

*A* = cross sectional area of borehole casing (excess head measured relative to initial groundwater level)

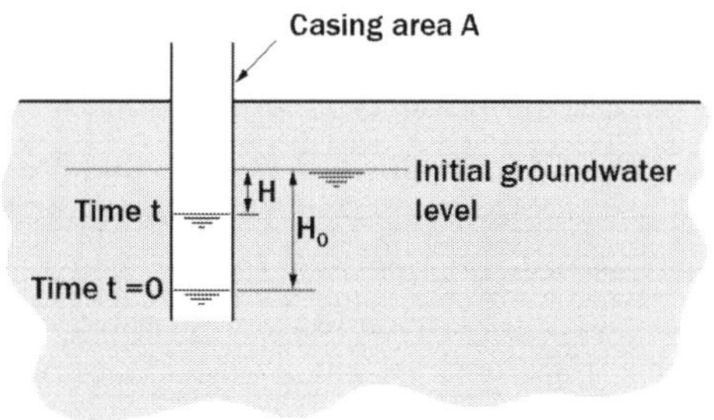


**Constant head test (outflow test shown)**

Water added or removed to change level in borehole and induce flow into or out of the ground

$H_c$ = constant excess head during test

*q* = flow rate

*A* = cross sectional area of borehole casing (excess head measured relative to initial ground water level)

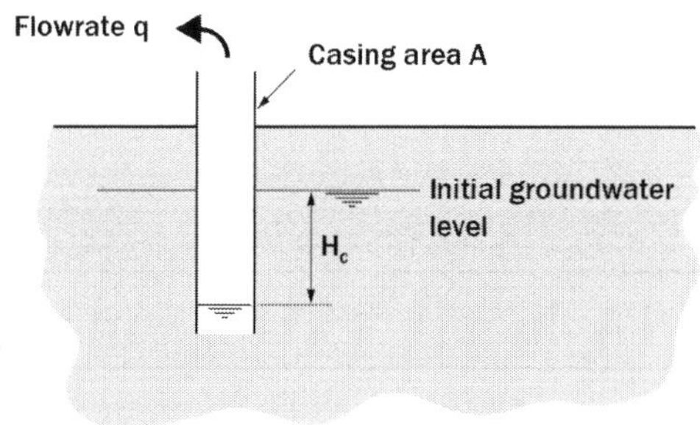


Tests and analysis are described in BS EN ISO 22282-2:2012.

## 5.3.3 Falling, rising and constant head tests in wells, standpipes and piezometers

These are tests carried out in wells, standpipes or piezometers installed in boreholes. The tests are described in detail in BS EN ISO 22282-2:2012. Provided the installations have been carried out in accordance with good practice these tests can give more reliable results than the same type of tests carried out in boreholes during drilling.

The tests are carried out in exactly the same way as for tests in boreholes (Box 5.3), but prior to testing the installation should be purged and developed to make good hydraulic connection between the response zone and the soil. It is possible to repeat tests several times or carry out several types of test on one piezometer. The following factors affect permeability estimates from tests in wells, standpipes and piezometers:

1. Development of the installation (typically by purging or pumping) is essential prior to testing.
2. Short piezometer screen lengths or finely screened piezometer tips can restrict flow, leading to underestimation of permeability in medium to high permeability soils.
3. Rising head or constant head outflow tests (as in Box 5.3) are preferable to falling head or constant head inflow tests. Clogging of the response zone may occur during falling head and inflow tests if the water added contains colloidal particles or gas bubbles.
4. Even though these tests are likely to give better results than borehole tests, because only a small volume of soil is tested, the results tend to be less reliable than from well pumping tests.

## 5.3.4 Packer tests

These *in situ* tests are carried out in unlined boreholes in rock (Box 5.4), and involve measuring the rate at which water can be pumped into or out of a section of borehole isolated between inflatable packers (double packer test) or between a packer and the bottom of the borehole (single packer test). Methods of packer testing are described in BS EN ISO 22282-3:2012, while methods of analysis are discussed by Houlsby (1976), and Wild and Money (1986).

In stable unlined rotary boreholes packer tests can be carried out at various depths, and may be useful in identifying fissured zones in rock. However, interpretation of the permeability values requires care because groundwater flow may be dominated by flow through fissures (see Walthall and Campbell, 1986, and Quinones-Rozo, 2010).). The following factors affect permeability estimates from packer tests in boreholes:

1 The measured permeability will be affected if drilling mud or debris block pores and fissures; this can lead to underestimates of permeability. On the other hand, drilling may have scoured and opened up infilled fissures, which can lead to overestimates of permeability.
2 Leaky or poorly sealed packers will result in excessive inflows and overestimates of permeability.
3 Injection pressures for pump-in tests have to be limited to avoid hydraulic fracturing or uplift of the ground. In very permeable zones the injection flow rate could be so large that the injection pressure cannot be maintained during the test.
4 The test section may not be representative of the rock mass in terms of fracture spacing, orientation and tightness. Fissure permeability can be greatly influenced by stress redistribution around the borehole.

*Box 5.4 Packer test*

Parameters to be recorded include the flow rate *Q*, applied head *H*, borehole radius *r* and length of the test section *L*.

*H* is defined as the head of water in the test section measured relative to initial groundwater level (for a pump-in test this is equal to the head shown on the supply pump pressure gauge, plus the depth from the gauge to the initial groundwater level, minus any head losses in the injection pipework).

Tests and analysis are described in BS EN ISO 22282-3:2012.

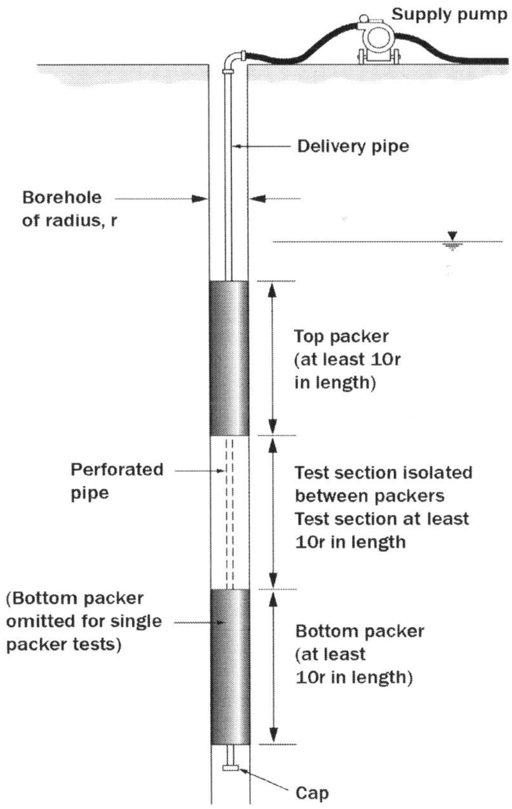

*General arrangement of double packer test in borehole*

## 5.3.5 Particle size analysis

See also

A1 Datasheet 4 Prugh method

Empirical methods are available to give approximate estimates of soil permeability based on the particle size distribution (PSD) curve from laboratory tests of disturbed soil samples (Box 5.5). The most common method is Hazen's formula (developed for uniform sands), which relates permeability $k$ in m/s to the $D_{10}$ particle size in mm (from the PSD curve):

$$k = C(D_{10})^2 \tag{5.1}$$

where $C$ is a calibration factor, which may vary between 0.007 and 0.017.

To obtain very approximate values of permeability it is usually sufficient to use $C = 0.01$ in Equation 5.1. Hazen's formula was developed for uniform filter sands, and may give misleading answers if applied to other soil types.

An alternative technique, suitable for use in less uniform soils, such as sandy gravel, is the Prugh method (Powers *et al*, 2007), which uses the $D_{50}$ particle size, uniformity coefficient $U$ (where $U = D_{60}/D_{10}$) and the relative density of the soil to estimate permeability from the three graphs of Datasheet 4 (Appendix A1), interpolating as necessary.

***Box 5.5 Particle size analysis of samples from boreholes***

$D_{10}$ = sieve aperture through which 10 per cent of a soil sample will pass (mm)

$D_{50}$ = sieve aperture through which 50 per cent of a soil sample will pass (mm)

$D_{60}$ = sieve aperture through which 60 per cent of a soil sample will pass (mm)

$U$ = uniformity coefficient = $D_{60}/D_{10}$

If more than 10 per cent of fine particles (<0.063 mm) are present, a sedimentation test is usually carried out to determine size distribution of fine particles.

Test methods are defined in BS EN ISO 17892-2:2014.

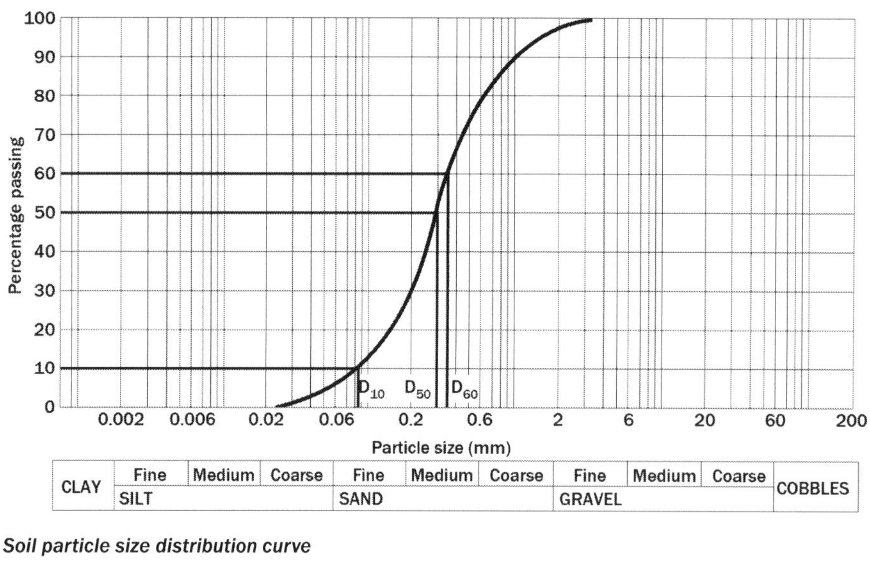


***Soil particle size distribution curve***

The following factors affect permeability estimates from particle size analysis:

1 When bulk or disturbed samples are taken from below water in a borehole, the samples obtained are unlikely to be representative of the deposit – in particular the finer particles will be washed out (known as loss of fines). This can cause permeabilities calculated from the PSD curves to be overestimates. Loss of fines is especially a problem for bulk samples taken from the drilling tool or shell. This can be minimised by placing the whole contents of the shell into a tank and allowing the fines to settle before decanting the water, but in practice this is rarely done. Loss of fines is usually less severe for tube samples (including SPT samples), and these may give more reliable samples for PSD testing of sands and finer soils.

2 If the *in situ* soil has significant fabric or layering, this can be destroyed during sampling and test specimen preparation. Permeability estimates from the resulting homogenised sample will be unrepresentative of the *in situ* permeability, which depends on soil fabric and may be significantly different in the vertical and horizontal directions.

3 The empirical rules for estimation of permeability are based on granular soils with relatively small proportions of fine (silt and clay) particles. It is inappropriate to use these methods for soils containing more than about 10 to 20 per cent of fine (ie silt and clay-sized) particles. *In situ* methods should be considered instead.

## 5.4 KEY REFERENCES


CLAYTON, C R I, MATTHEWS, M C and SIMONS, N E (1995) Site investigation: a handbook for engineers, second edition, Wiley-Blackwell, London (ISBN: 978-0-63202-908-2)

KRUSEMAN, G P and DE RIDDER, N A (1990) *Analysis and evaluation of pumping test data, second edition*, Publication 47, International Institute for Land Reclamation and Improvement, Wageningen, The Netherlands (ISBN: 978-9-07075-420-4)

SITE INVESTIGATION STEERING GROUP (1993) *Site investigation in construction. Part 1 Without investigation ground is a hazard*, Institution of Civil Engineers, London (ISBN: 978-0-72773-983-4)

SITE INVESTIGATION STEERING GROUP (2013) *Effective site investigation (Site investigation in construction series)*, Thomas Telford, London (ISBN: 978-0-72773-505-8)


### Standards


BS 5930:2015 *Code of practice for ground investigations*

BS EN 1997-2:2007 *Eurocode 7: Geotechnical design. Part 2: Ground investigation and testing*

BS EN ISO 22282-1:2012 *Geotechnical investigation and testing. Geohydraulic testing. General rules*

BS EN ISO 22282-2:2012 *Geotechnical investigation and testing. Geohydraulic testing. Water permeability tests in a borehole using open systems*

BS EN ISO 22282-3:2012 *Geotechnical investigation and testing. Geohydraulic testing. Water pressure tests in rock*

BS EN ISO 22282-4:2012 *Geotechnical investigation and testing. Geohydraulic testing. Pumping tests*

BS EN ISO 22282-5:2012 *Geotechnical investigation and testing. Geohydraulic testing. Infiltrometer tests*

BS EN ISO 22282-6:2012 *Geotechnical investigation and testing. Geohydraulic testing. Water permeability tests in a borehole using closed systems*

BS ISO 14686:2003 *Hydrometric determinations. Pumping tests for water wells. Considerations and guidelines for design, performance and use*

# 6 Analysis and design

The idea of the way in which the ground and groundwater will behave is called the conceptual model. A conceptual model is the basis for all designs, whether carried out by supercomputer or scribbled on the back of an envelope. This chapter considers the factors important to the development of conceptual models and discusses selection of permeability values and methods of numerical modelling. Sections 6.2 to 6.6 give methods for the estimation of steady-state flow rate, design of wells and filters, time–drawdown relationships, drawdown patterns around wells, and settlements from groundwater control activities. Chapter 7 presents some case studies, which illustrate the transition from design to practice. Figure 6.1 is a flowchart showing some of the principal stages of analysis and design.

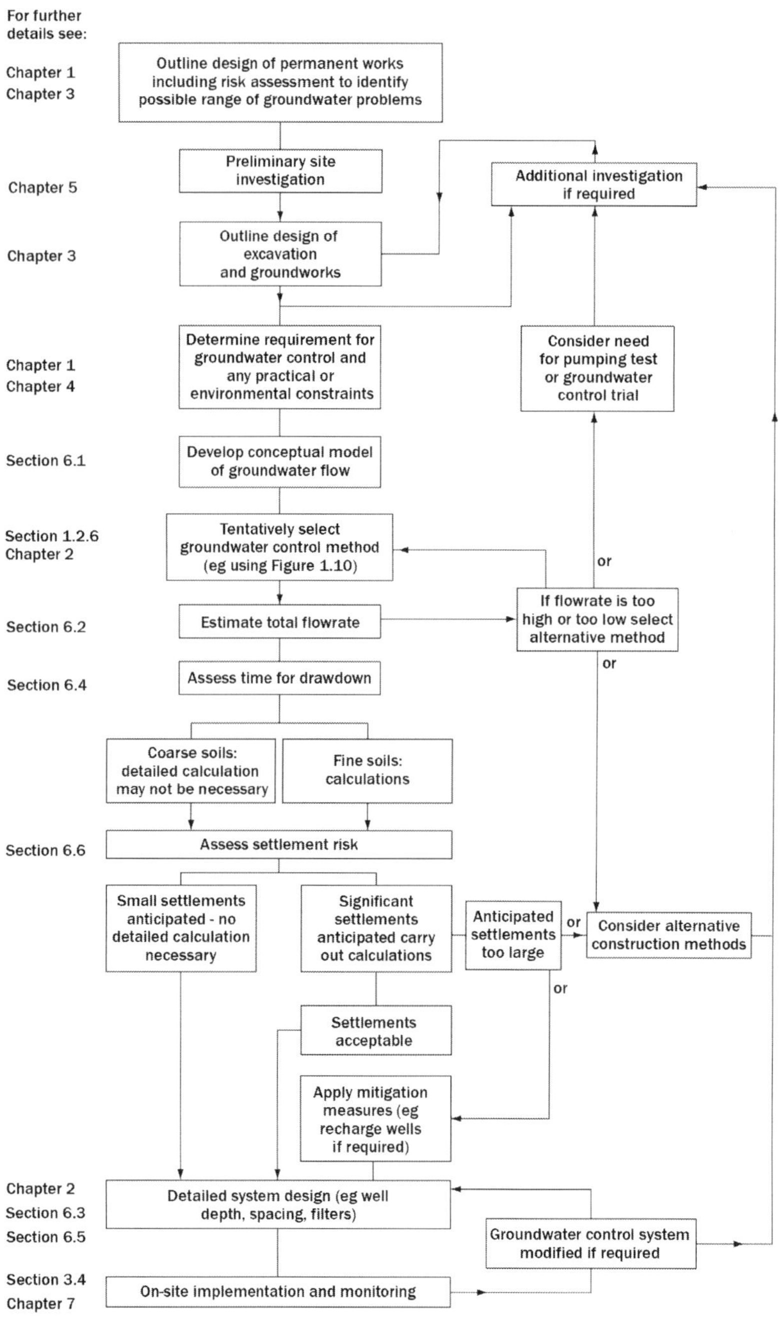


*Figure 6.1 Principal stages in the analysis and design of groundwater control systems*

See also

Chapter 7 Design to practice

In the UK, the requirements for geotechnical design are set out in *Eurocode 7: Geotechnical design – Part 1: General rules* (BS EN 1997-1:2004) (EC7), which includes sections on dewatering and on hydraulic failure of excavations (eg by base heave or piping, see Section 1.2.5). Further discussion of EC7 is given in Driscoll *et al* (2008).

# 6.1 GROUNDWATER MODELLING AND SELECTION OF DESIGN PARAMETERS

## 6.1.1 Modelling of groundwater flow

*When analysing a groundwater control system, the concept of how water will flow in the ground must be correct. If it is not correct, even though permeability is carefully selected and calculations are meticulously carried out, the results are likely to be wildly unrealistic.*

The conceptual model comprises the geometry and boundary conditions of groundwater flow and the permeability values of the various strata. A method of analysis based on the direct application of a standard theoretical solution (eg from a text book) is called an analytical model. A solution, normally done on computer, which breaks down the problem into discrete parts and solves the whole (often by iteration) is called a numerical model (Section 6.1.4). Table 6.1 lists some of the key information required to develop a conceptual model.

The selection of the modelling approach should be proportionate to the complexity of the ground conditions. For example, in a simple geological setting with anticipated low pumping rates then a good conceptual model and simple analytical equations may be all that is required. Conversely, long-term pumping at high flow rates in a layered aquifer system will still require a strong conceptual model but may also require more complex numerical modelling.

See also

Chapter 5 Site investigation

6.1.4 Numerical modelling

Selection of permeability values is recognised as important for the design of groundwater control schemes, as poor estimates can result in large errors. It is not so well understood that the boundary conditions, which make up the conceptual model, can dramatically affect calculations; care must be taken in assessing those to be used in design. It is essential to have site specific information about the potential groundwater problems and adequate data for design (Chapter 5). Often uncertainties will remain; a parametric or sensitivity analysis (Box 6.1) may be required to determine the effects of different boundary conditions. Box 6.2 shows a case study demonstrating the effect of ambiguous boundary conditions on flow rate calculations.

*In geotechnical engineering it is usually necessary to make simplifications to arrive at a conceptual model of a real situation, which is amenable to analysis. The secret of success is not to ignore any factor, which could destroy the applicability of the conceptual model adopted.*

*Box 6.1 Sensitivity and parametric analyses*

**Sensitivity analysis**

A way of determining how the results of the model will alter if one parameter is varied. For example, *"How will the calculated flow rate change if the permeability is different from the expected value?"* It can clarify whether a given parameter is known within sufficient limits or whether additional investigation is required.

**Parametric analysis**

A broader form of sensitivity analysis to answer the question *"What parameters are important to the model?"* It involves modelling a range of values for several parameters to determine which parameters have the greatest effect and may therefore need further investigation.

*Box 6.2 Case study of the effect of boundary conditions on the design of a dewatering system*

An excavation was to be made through a high permeability gravel stratum into low permeability silty sand. Boreholes showed these strata were separated by a thin, possibly discontinuous clay layer. A sheet-pile cut-off wall was to be used to exclude groundwater in the gravels from the excavation, so the proposed wellpoint system was to pump from the silty sand only. Assumptions made about the thin clay layer have a dramatic effect on the analysis. If the clay layer is assumed to be continuous and impermeable, the gravel is not linked to the sand (see figures), and flow will be from the sand only, with a distance of influence of approximately 50 m. Equivalent well calculations predict a flow rate of approximately 4 l/s.

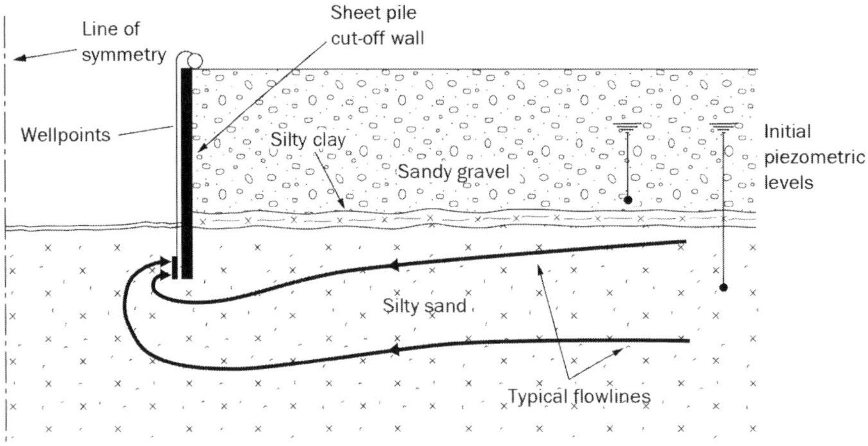


**Continuous clay layer: flow is from sand only**

However, if the clay layer is assumed to be so thin that it is either discontinuous or effectively permeable, the main groundwater flow will be from the overlying permeable gravels, which act as a close source of recharge (see figure). Flow rates are then much higher: flownet analyses predict approximately 50 l/s.

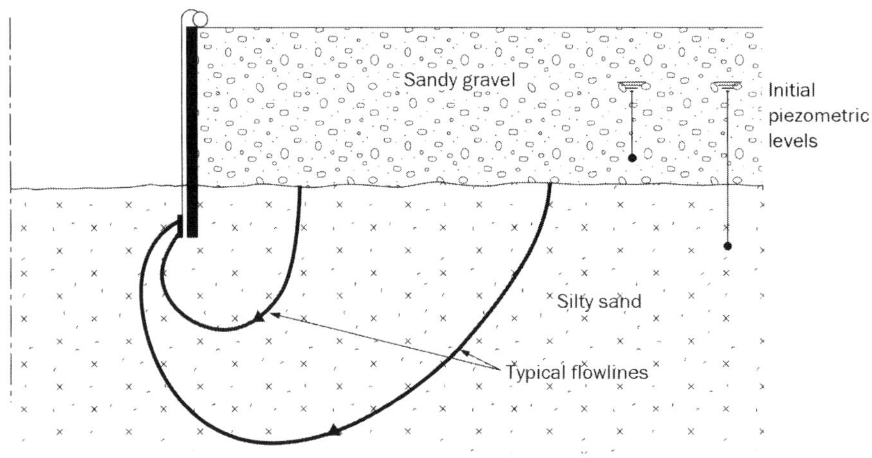


**Clay layer absent: flow from gravel dominates**

This case study shows that small geological features, such as the clay layer, can affect groundwater flows. Even if more investigation boreholes had been drilled, some uncertainty would have remained, and both cases should have been modelled. In fact, when the site was dewatered, the flow rate was 7 l/s, indicating that conditions lay somewhere between the two cases, but closer to the first than the second.

*Table 6.1 Key components of a conceptual model for groundwater control design*

| Aspect | Comments |
|---|---|
| Aquifer depth and thickness | What are the depth and thickness of the permeable strata?<br>Are the depth and thickness constant or do they vary across the area affected by the dewatering? |
| Aquifer boundaries | Are the boundaries known?<br>Is the aquifer bounded by impermeable layers or more permeable strata? |
| Source of recharge and distance of influence | Can the water come from a nearby source (eg a river) or purely from water stored in the ground?<br>Can the approximate distance of influence be estimated? |
| Initial groundwater levels or pore water pressures | What are the groundwater levels or pore water pressures in relation to the depth of excavation?<br>Are they influenced by seasonal, tidal or other effects? |
| Hydrogeological parameters | Are there reliable estimates of permeability and storage coefficients?<br>Are the parameters likely to vary within the area affected by the dewatering?<br>Are the parameters different for each soil layer?<br>Is the permeability likely to be isotropic or anisotropic? |
| Environmental constraints | Are any specific limits or conditions on pumping or discharge likely to be imposed by the regulatory authorities? |
| Geometry of excavation | Required to determine the necessary depth and extent of drawdown.<br>Will any cut-off walls be used and how will this affect the groundwater flow? |
| Groundwater control technique | Formulation of an appropriate model may be easier if an initial guess is made of the likely pumping technique (eg by using Figure 1.10). |

In some cases, especially on larger projects or where the presence or risk of groundwater contamination has been identified at site investigation stage (see Section 4.1.5) groundwater quality may need to be modelled, potentially including the assessment of:

- The mobilisation and movement towards the groundwater control system from contamination sources (such as light non-aqueous phase liquids (LNAPLs) or contaminated soils).
- The risk to groundwater quality from hazardous and non-hazardous substances from any nearby groundwater activity (such as a soakaway or artificial recharge).

These approaches may require different methods or complexity of modelling for either solute transport or the risk from pollution. Specialist advice from a suitably qualified and experienced specialist may be required.

## 6.1.2 Concepts of groundwater flow

The factors which influence the development of an appropriate conceptual model are described briefly as follows.

### Unconfined and confined aquifers

The distinction between unconfined and confined aquifers is important. An unconfined (or water table) aquifer is characterised by the top of the aquifer being open to atmosphere, and the water table or phreatic surface (ie the line of zero pore water pressure) being below ground level (Box 6.3). In comparison, a confined aquifer is that which is overlain (or confined) by an effectively impermeable stratum (eg a clay layer), known as an aquiclude, and the piezometric level is above the top of the aquifer, so the aquifer is saturated throughout (Box 6.3). If a borehole is drilled into the aquiclude, it will not encounter flowing groundwater until it penetrates the aquifer, at which point water will enter the borehole and, over a period of time, rise up to the piezometric level. Confined aquifers are sometimes known as sub-artesian aquifers. If the piezometric level is above ground level, wells penetrating the aquifer may overflow naturally (see Box 1.1).

Leaky aquifers are those where the main permeable aquifer is overlain by a confining layer which is 'semi-permeable' (eg silt or clay containing layers of fine sand), known as an aquitard. When the aquifer is pumped and piezometric levels are lowered, water will drain slowly into the aquifer from the aquitard and will contribute to the flow (see Bear, 1979).

*Box 6.3 Unconfined and confined aquifers*

**Unconfined aquifer:**

- top of the aquifer is open to the atmosphere
- aquifer is saturated below the water table and may be unsaturated above
- water table (or phreatic surface) is the level at which pore water pressures are zero.

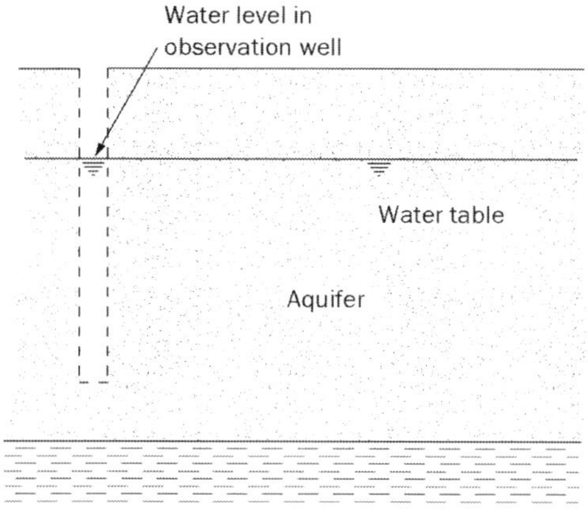


**Confined aquifer:**

- aquifer is overlain by an effectively impermeable stratum and the whole of the aquifer is saturated, with pore water pressures greater than zero
- pore water pressures are described by the piezometric level (the level shown in piezometers installed in the aquifer).

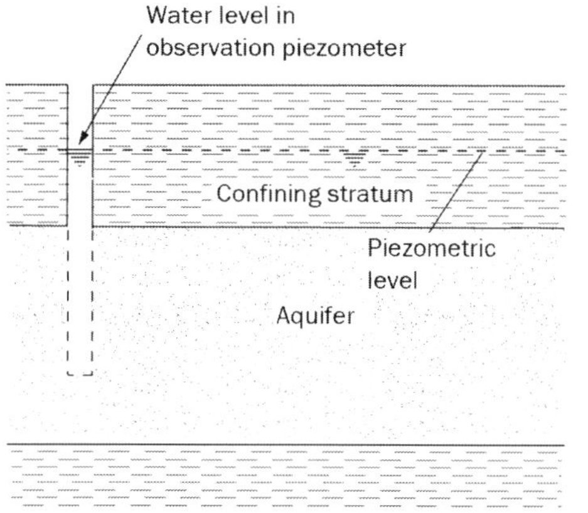


The permeability (or hydraulic conductivity) of the saturated aquifer is given the symbol, $k$. Transmissivity, $T$, is the product of multiplying the permeability by the saturated aquifer thickness, $D$: $T = kD$. For an unconfined aquifer, $T$ will reduce significantly during pumping as the water table is lowered and the saturated thickness reduces. This is one of the factors that makes analysis of unconfined aquifers more complex than analysis of confined aquifers.

## Groundwater storage

**See also**
1.2.6 Objectives of groundwater control
5.3.1 Well pumping tests

When a well pumps from an aquifer, unless there is a close source of recharge (eg a river in hydraulic connection with the aquifer), the pumped water will come from water stored in the soil pores. Pumping lowers the water table or piezometric level and releases water from storage. The aquifer storage coefficient, $S$, is a measure of how much pore water can drain out of a soil by gravity. $S$ is defined as the volume of water released from storage per unit area per unit reduction in head; it is a dimensionless ratio (see Bear, 1979).

In an unconfined aquifer, as the water table is lowered, water drains out of soil pores above the phreatic surface. Coarse-grained soils, such as sands and gravels, desaturate above the water table; fine-grained soils, such as silty sands, do not drain so freely, and may remain saturated above the lowered water table. In such soils groundwater control should be thought of as pore water pressure control rather than dewatering (see Section 1.2.6). An unconfined gravel stratum may have a storage coefficient as large as 0.1 to 0.2, but for a silty sand the storage coefficient will be significantly lower, as much of the pore water is retained by surface tension forces.

In a confined aquifer groundwater control lowers the piezometric level, and the pore water pressures are reduced, but the aquifer does not desaturate unless the piezometric level drops below the top of the aquifer. Because the aquifer remains saturated, water is released only as a result of the compression of the aquifer and expansion of the pore water, so the storage coefficient is much less. *S* may be of the order of 0.0005 to 0.001 or even less. Over time, pumping from the aquifer may result in reduction of pore water pressures (and hence consolidation) in the aquiclude (or the aquitard in a leaky aquifer); the aquifer is said to be underdraining the aquiclude. If the piezometric level drops below the top of the aquifer, unconfined conditions may develop locally, particularly around wells, and the aquifer is described as mixed (ie partly confined and partly unconfined). *S* can normally be estimated from an appropriately analysed pumping test (Section 5.3.1) as described by Kruseman and De Ridder (1990).

## Plane and radial flow

Flow to a groundwater control system is three-dimensional. However, to simplify calculations, flow is often analysed as two-dimensional in cross-section. If the flow regime cannot easily be simplified to two dimensions, numerical analysis in three dimensions may be required (Section 6.1.4).

For two-dimensional analyses it is important to assess whether the groundwater flow is best modelled as plane or radial. As discussed in Section 6.2, for plane flow the system may be modelled as an equivalent slot; for radial flow the system may be modelled as an equivalent well. The flow pattern (Box 6.4) depends on both the aquifer recharge conditions and the layout of the dewatering system. These define the hydraulic boundaries of the analytical model.

*Box 6.4 Plane and radial groundwater flow*

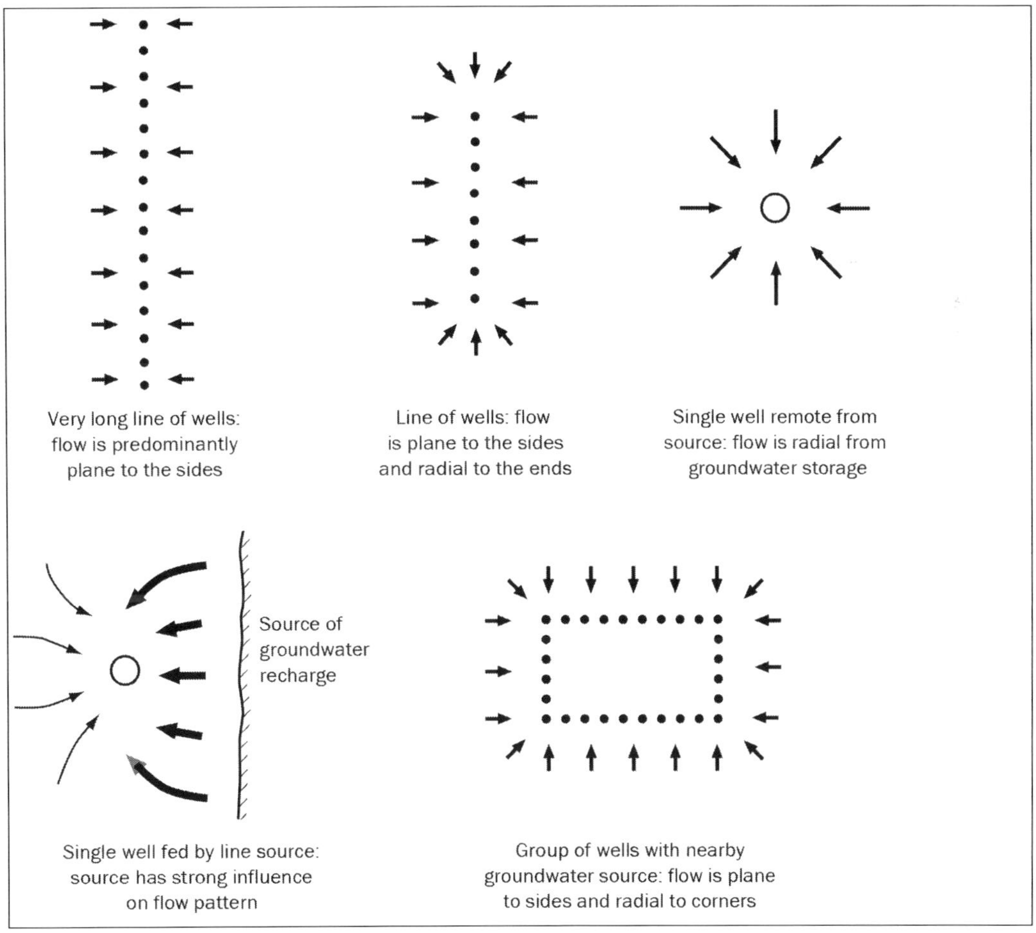

Plane flow is likely when:

- the system consists of long lines of wells (eg for a linear trench excavation)
- the system consists of a large ring of wells and a groundwater source is very close (eg permeable gravels around a cofferdam).

Radial flow is likely when:

- the system consists of a ring of wells with no nearby groundwater source (flow is purely from groundwater storage).

Flow to a system can be plane and radial in different areas, eg a line of wells of finite length may have plane flow to the sides and radial flow to the ends. This is discussed further in Section 6.2.1.

## Sources of groundwater recharge and other boundary conditions

If there is a specific feature linked to the aquifer, which can act as a source of groundwater (Figure 6.2), the distance between the dewatering wells and the source will have a major effect on flow. In simple terms, the closer the source of recharge is to the dewatering system, the greater the pumped flow rate will be. Barrier boundaries may also exist, which block groundwater flows and tend to reduce pumped flow rates. Just because a river runs close to a site, there is no certainty that it is in hydraulic connection with the aquifer and will act as a source of recharge, see Case study 7.14. Slow-moving rivers can have thick silt beds, which separate the river from the aquifer. The potential sources of recharge should be assessed during the desk study and site reconnaissance stages of site investigation (Section 5.2). Analysis of pumping test results may also allow identification of boundary conditions (see Kruseman and De Ridder, 1990).

If there is no specific groundwater source nearby, all flow to the system will be from groundwater storage in the soil pores of a zone around the system, which grows with time. In analysis, the boundary of this zone is defined by a theoretical term called the distance of influence for plane flow and radius of influence for radial flow.

## Distance of influence and radius of influence

**See also**
5.3.1 Well pumping tests

The distance (or radius) of influence is defined as the distance from a well or a system of wells to the point at which drawdown is just equal to zero. Pumped flow rate is affected by the distance of influence – all other factors being equal, a smaller distance of influence will result in a greater flow rate. Distance of influence is not a constant – theoretically it is zero at the instant pumping starts and gradually increases with time while the pumping continues (Box 6.5). In reality, a steady-state distance may never exist, but as time passes, the distance of influence will increase at an ever diminishing rate and can approach a quasi steady-state. If there is a source of groundwater recharge nearby, a true steady-state distance of influence may exist (being the fixed distance between the wells and the source).

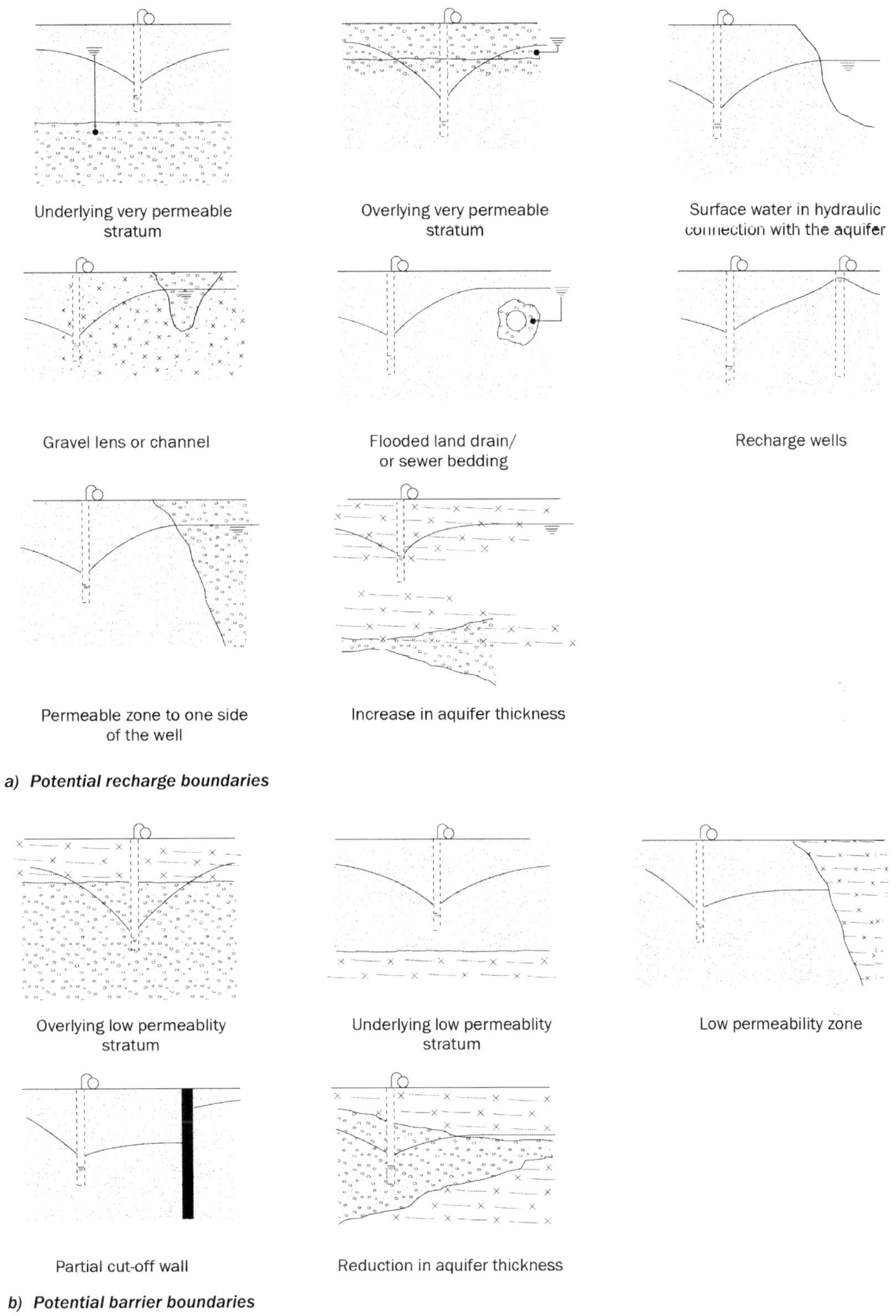


*Figure 6.2 Potential aquifer boundary conditions*

*Box 6.5 Distance of influence*

A well or slot is installed into an unconfined aquifer with groundwater levels initially constant. Just after pumping begins, the water level outside the well will be hardly affected by pumping. But, as time passes, the drawdown propagates away from the well and the distance of influence increases.

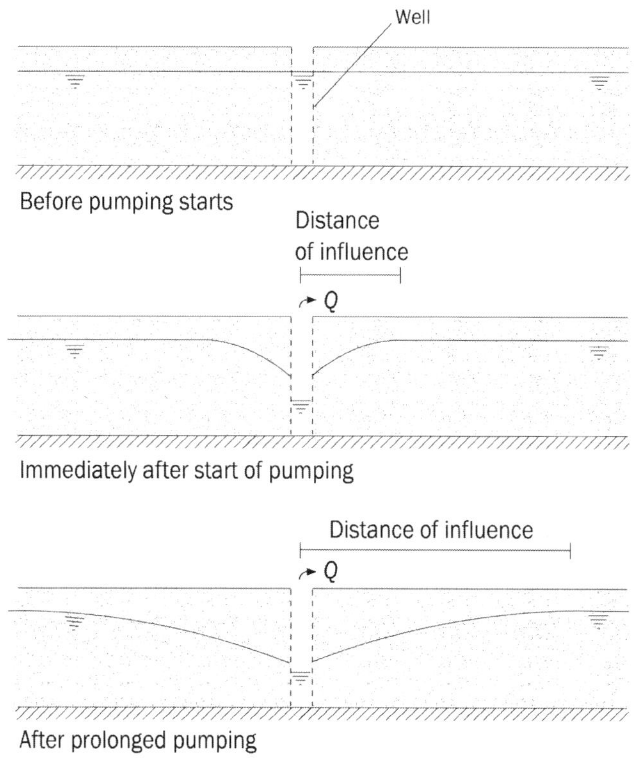


For plane flow, the distance of influence has the symbol $L_o$. If the flow is radial, the symbol for the radius of influence is $R_o$. $L_o$ or $R_o$ can only be determined from appropriately analysed pumping tests (Section 5.3.1) or groundwater control trials. In the absence of such data, the following mathematical expressions can be used in confined aquifers to estimate $L_o$ (from Powrie and Preene, 1994a) and $R_o$ (Cooper and Jacob, 1946):

Plane flow: $$L_o = \sqrt{\frac{12kDt}{S}} \tag{6.1}$$

Radial flow: $$R_o = \sqrt{\frac{2.25kDt}{S}} \tag{6.2}$$

where $t$ is the time since pumping started, $S$ is the aquifer storage coefficient, $D$ is the aquifer thickness and $k$ is the soil permeability. For relatively compressible soils (where the water released from storage is predominantly from compression of the aquifer, and the amount released by the expansion of pore water is relatively small), these equations can be expressed approximately as:

Plane flow: $$L_o = \sqrt{\frac{12kE'_o t}{\gamma_w}} \tag{6.3}$$

Radial flow: $$R_o = \sqrt{\frac{2.25kE'_o t}{\gamma_w}} \tag{6.4}$$

where $E'_o$ is the stiffness of the soil in one-dimensional compression (estimation of $E'_o$ is discussed in Section 6.6.2 and Table 6.4) and $\gamma_w$ is the unit weight of water. Because $L_o$ and $R_o$ are proportional to $\sqrt{(kt)}$, at a given time the distance of influence will be greater for a high permeability soil than for one of lower permeability. $L_o$ and $R_o$ are also inversely proportional to the aquifer storage coefficient, $S$, so at a given time the distance of influence will be greater in an aquifer with a small $S$ compared with an aquifer with a large $S$. Equations 6.1 and 6.2 were derived for confined aquifers but are sometimes applied in unconfined conditions by substituting the unconfined values of $S$. In confined aquifers when S is small, drawdown response to pumping and the rate of expansion of the distance of influence will be much quicker than in an unconfined aquifer where $S$ will be much greater.

See also
Table 6.4 Estimation of soil stiffness

*It is important that realistic values of distance of influence are used in calculations. For example, a calculated $R_0$ of several kilometres could not sensibly be used in an aquifer of limited areal extent. At the other end of the scale, in a low permeability silty sand, $R_0$ may be calculated as only a few metres, which indicates that a system relying on the interaction of drawdown between wells may not be effective. In reality very large or very small calculated distances of influence usually indicate that either unrealistic permeability values have been used, or pumping tests have been analysed using inappropriate boundary conditions.*

Even if the long-term $R_0$ were known, it should not be used in design because it would predict a relatively low flow rate. A system designed on that basis might not be able to cope with higher flow rates in the short term – when $R_0$ is small soon after pumping starts. Equally, if a system were designed on the basis of a very small $R_0$, a much larger flow rate would be predicted, which would suggest unrealistically large pumping rates (Section 6.2.1 discusses the use of Sichardt's empirical formula, Equation 6.8, to estimate $R_0$ and $L_0$ for steady-state flow rate calculations). Also, if a source of groundwater recharge (Figure 6.2) is nearby, the distance or radius of influence will be affected by it and Equations 6.1 to 6.4 may not apply.

## Aquifer depth and partial penetration

If the aquifer is of finite thickness, the wells may be installed to penetrate right through the aquifer down to the underlying stratum, this is called a fully penetrating system. If the aquifer is very thick, this may not be cost-effective, and shallower partially penetrating wells may be used (Figure 6.3).

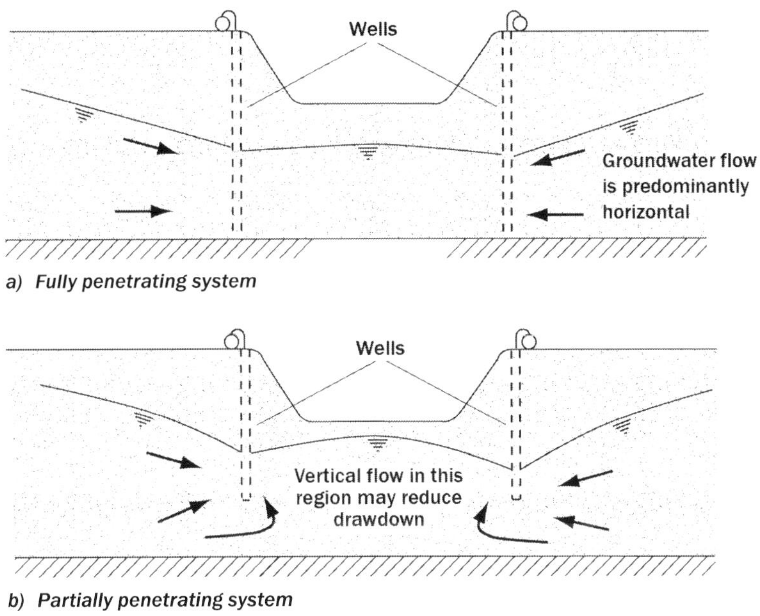


*Figure 6.3 Fully and partially penetrating systems*

Flow through the aquifer to fully penetrating wells is predominantly horizontal; an assumption of horizontal flow is the basis of many methods of analysis. Partially penetrating wells will introduce vertical flow in the aquifer and will alter the drawdown pattern around the system. It is useful if the type of penetration (full or partial) is identified when the conceptual model is formulated; different analyses may be required for each case.

## Vertical groundwater flow and underdrainage

A partially penetrating system is one example of when there is a vertical (as opposed to purely horizontal) component of groundwater flow. Another is when a dewatering system is used in combination with a physical cut-off wall to reduce pumped flow rates (Figure 6.4a).

See also
Case study 7.1

Soil stratification also affects the flow pattern. Figure 6.4b shows an example of a highly permeable layer of gravel at depth which will feed water upwards through the overlying sand. This can cause significant problems for dewatering systems and it may be necessary to increase the well depth to penetrate the deep gravel layer. If the wells were extended to pump directly from the gravel layer, lowering the piezometric level in that stratum, the shallow sands might drain downwards into the deep permeable stratum. This is known as underdrainage (see Case study 7.1). Where there is a permeable stratum at depth, underdrainage can be an efficient way of dewatering shallow strata.

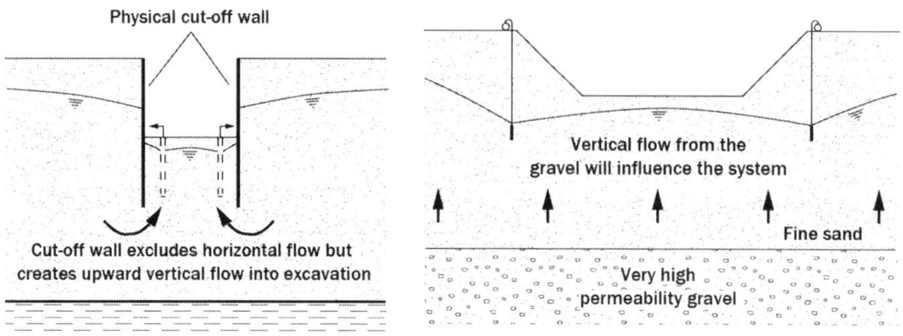


*a) Cut-off wall* *b) Very permeable stratum at depth*

*Figure 6.4 Vertical groundwater flow*

It is important to identify the potential for vertical flow, because then the vertical and the horizontal permeability, $k_v$ and $k_h$ respectively, of the soil will be relevant, rather than just $k_h$ for purely horizontal flow.

## 6.1.3 Selection of permeability

See also

| | |
|---|---|
| Table 1.1 | Permeability values |
| 5.3 | Permeability testing |
| Box 6.1 | Sensitivity analysis |

Section 5.3 describes the difficulty of obtaining representative permeability values because of limitations of the test methods available (see Table 6.2) and the natural variations and anisotropies of the ground. These include:

- existence of different strata of variable thickness and extent
- presence of fissures or lenses
- soil fabric and structure.

*There are no easy solutions to this problem: uncertainty of permeability values is unavoidable.*

The permeability values used by designers have to represent the permeability of a large soil mass, with all its variability, rather than a discrete element or sample. The challenge is to select mass permeabilities, which reflect the dominant characteristics of the aquifer. A common approach is to carry out a sensitivity analysis (Box 6.1), repeating the analysis for the possible range of permeability, in order to assess the impact of the uncertainty in permeability. Box 6.6 shows a permeability sensitivity analysis for a flow rate calculation.

*Table 6.2 Tentative guide to reliability of permeability estimates from various methods*

| Method | Notes | Reliability |
|---|---|---|
| **Groundwater control trials** | Appropriate for large-scale works or where the observational method is being used (Section 7.2)<br>Costs high but can be offset against main works | Good if appropriately analysed |
| **Pumping tests** (Section 5.3.1) | Test a large volume of soil<br>Provide information on well yields, water chemistry and distance of influence<br>Costs high to moderate depending on complexity (simple tests can sometimes be very useful) | Good if appropriately analysed<br>May be difficult to carry out and analyse in fine-grained soils (eg silt) or if there are different strata with significant variations in permeability (eg gravel over fine sand) |
| **Inverse numerical modelling** | Uses groundwater monitoring data (eg piezometer readings) to back-analyse permeability<br>Costs low to moderate | Good if adequate groundwater data is available and is appropriately analysed |
| **Tests in boreholes** (Section 5.3.2) | Only a small volume of soil tested. Affected by soil disturbance from drilling | |
| Falling head | Very prone to clogging of borehole<br>Costs very low | Very poor |
| Rising head | Prone to clogging or loosening of borehole base<br>Costs very low | Poor to moderate<br>Better results in coarser, less silty soils |
| Constant head | Inflow tests prone to clogging<br>Costs low | Poor to moderate<br>Better results in coarser, less silty soils |
| Packer test (Section 5.3.4) | Normally carried out in rock<br>Results are hugely influenced by fissure network<br>Costs low to moderate | Poor to good<br>Can confirm presence of fissures depending on fissure spacing |
| **Tests in piezometers and standpipes** (Section 5.3.3) | Only a small volume of soil tested; dependent on the design and quality of piezometer installation and on any soil disturbance | |
| Falling head | Prone to clogging<br>Costs very low | Very poor |
| Rising head | Costs very low | Poor to moderate<br>Better results in coarser, less silty soils |
| Constant head | Inflow tests prone to clogging<br>Costs low | Poor to moderate<br>Better results in coarser, less silty soils |
| **Specialist *in situ* tests**, eg piezocone, *in situ* permeameter | Can identify small stratigraphic changes and provide permeability profile with depth<br>Costs moderate | Can be good in fine-grained soils (silts and clays)<br>Can be difficult to use in coarser soils or weak rocks |
| **Laboratory tests** | Only a small volume of soil tested; sample disturbance can affect results | |
| Particle size (PSD) analysis of bulk samples (Section 5.3.5) | Loss of fines during sampling may lead to overestimates of permeability<br>Costs very low to low | Very poor, especially in laminated or structured soils |
| Particle size analysis of tube samples (Section 5.3.5) | Not representative in structured soils or if silt and clay content is more than about 10 to 20 per cent<br>Costs very low to low | Moderate to good in uniform sands with low silt and clay content; poor in laminated or structured soils |
| Permeameter testing, eg triaxial cell, Rowe consolidation cell, oedometer consolidation cell | Soil fabric and structure means sample size affects results: smaller samples tend to underestimate *in situ* permeability (Rowe, 1972)<br>Costs low to moderate | Good in clays and some silts where minimally disturbed samples, large enough to be representative, can be obtained |

#### *Box 6.6 Example of permeability sensitivity analysis applied to a flow rate calculation*

Groundwater control for a shaft 25 m deep using fully penetrating wells in a confined medium sand aquifer, analysed by the equivalent well method (see Section 6.2.1).

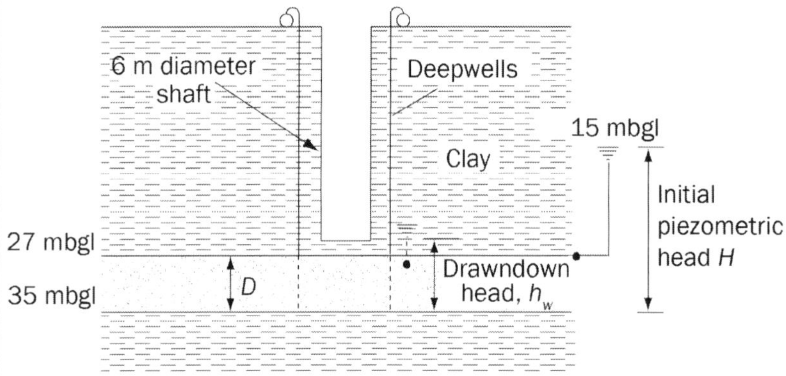


Flow rate, Q, calculated using Equation 6.5: $$Q = \frac{2\pi kD(H - h_w)}{\ln\left[\frac{R_o}{r_e}\right]} \tag{6.5}$$

| | |
|---|---|
| Aquifer thickness: | D = 35 – 27 = 8 m |
| Initial piezometric head: | H = 35 – 15 = 20 m |
| Drawndown piezometric head: | $h_w$ = 35 – 26 = 9 m |
| therefore drawdown is: | (H – $h_w$) = 20 – 9 = 11 m |
| Radius of influence: | $R_o$ = 500 m (assumed) |
| Equivalent radius of system: | $r_e$ = 6 m (based on wells 3 m outside shaft) |

Permeability *k* is estimated to be in the range $2 \times 10^{-4}$ to $5 \times 10^{-4}$ m/s.

Using these parameters *Q* is calculated as:

| Permeability k (m/s) | Flow rate Q (m³/s) | Flow rate Q (l/s) |
|---|---|---|
| $2 \times 10^{-4}$ | 0.025 | 25 |
| $3 \times 10^{-4}$ | 0.038 | 38 |
| $4 \times 10^{-4}$ | 0.050 | 50 |
| $5 \times 10^{-4}$ | 0.063 | 63 |

**Note**

If Sichardt's formula (Equation 6.9) is used to estimate $R_o$, the $R_o$ value will vary with permeability and this should be included in the analysis. In this case, for simplicity, $R_o$ was assumed to be constant.

In order to carry out a permeability sensitivity analysis, the probable range of permeability has to be determined from the available data. Table 5.4 lists the commonly used methods of permeability testing. Although permeability may vary across a site, very large variations in measured values may be because of limitations in the test techniques (see Section 5.3) and results may not reflect the properties of the ground. Judgement is needed to reduce the range of test results to a narrower range of probable permeability.

Assessment methods include:

- Comparing the permeability results with approximate permeability values based on visual descriptions of the soil (for example Table 1.1). This can allow very high or low test results to be discounted, eg a soil described as a sandy gravel is unlikely to have a permeability of $10^{-8}$ or $10^{-7}$ m/s as is sometimes reported by results of falling head tests in boreholes.
- Comparing the range of permeability values with the conceptual model. Is the permeability expected to vary with depth or across the site? For example, is there a wide range of values because two different strata have been tested and the results not differentiated?

- Differentiating the test results by test type. On a given site different test methods can give wildly differing results. Different methods can be classified as more or less reliable – some guidance is given in Table 6.2.

The permeability may be anisotropic (eg $k_v$ is different to $k_h$). The conceptual model will affect which permeability value is used. For example, a fully penetrating system may generate predominantly horizontal flow, so $k_h$ will be appropriate. If vertical flow is likely, $k_v$ may be needed as well. Even without the effects of clogging, sample disturbance etc the permeability value provided by many of the available test methods is neither $k_h$ nor $k_v$ but a theoretical equivalent isotropic permeability $\sqrt{(k_v k_h)}$.

## 6.1.4 Computer and numerical analysis

Advances in personal computer (PC) technology mean that computer modelling can be cost-effective for even very small groundwater control projects. PCs can be used in two ways:

1. To apply an analytical model which might previously have been carried out by hand (eg equivalent well flow rate calculations), using a spreadsheet with calculation routines usually written by the dewatering designer for each project (see Section 6.5.2). The advantage of using a PC and spreadsheet is that calculations can be repeated very easily to carry out a permeability sensitivity analysis or a parametric study (Box 6.1) of the effect of different boundary conditions.
2. To apply a numerical model, typically a groundwater modelling package written by a geotechnical software producer, which allows complex boundaries and geometries, not amenable to analytical solution, to be modelled. Such packages may also offer high quality graphical output (Box 6.7) and allow calculations to be easily repeated for parametric studies.

**See also**
Box 6.1 Sensitivity analysis

Detailed recommendations on the selection and use of numerical groundwater modelling packages are beyond the scope of this report. The different types of numerical models available are discussed in Chapter 8 of Holden *et al* (1998). Further background on modelling can be found in Anderson and Woessner (2015), and Chapter 14 of Fetter (2014). Advice on the validation and use of geotechnical software is given in the guide produced by the Association of Geotechnical and Geoenvironmental Specialists (Bond, 1994). Some specific notes on the application of numerical modelling to the design of groundwater control systems are given in the following section.

***Box 6.7 Example of graphical output from numerical model***

The figure shows the results of seepage analysis for the case shown in Figure 6.12. The distribution of groundwater head is described by the equipotentials, and flowlines are indicated by the Darcy velocity vectors.

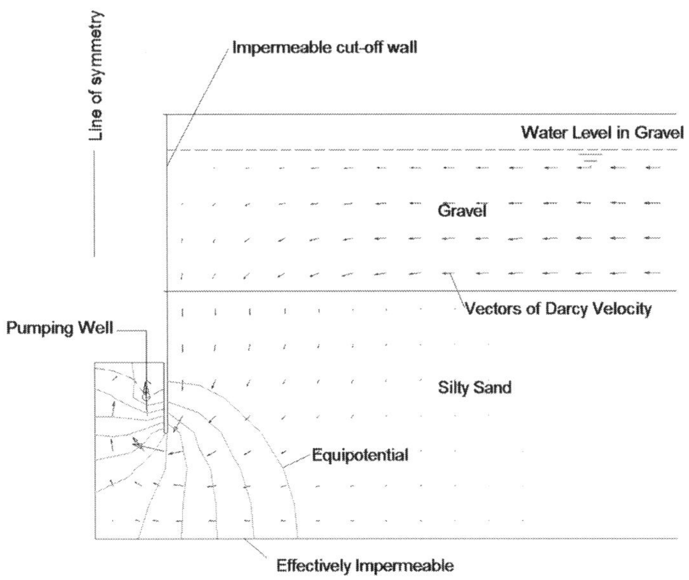

Numerical modelling consists of five stages:

1 Selection of software.
2 Setting up the model.
3 Calibration.
4 Prediction.
5 Refinement.

## Selection of software

Selection of the program will depend on the nature of the conceptual model. The following considerations are relevant (Bond, 1994):

- Will the software do what is needed?
- Will the software perform as claimed?
- Is the user qualified and ready to use it?

Whether the software can model the situation in hand is likely to be the deciding factor; some software can model three-dimensional flow, but many programs are intended for two-dimensional flow only.

## Setting up the model

To reduce errors at this stage Bond (1994) recommends:

- Never use a model that is more complex than necessary.
- Consider the results to be in error until they have passed rigorous scrutiny.
- Even if all checks indicate that the results are valid, it does not mean they are correct.
- Do not attempt to model detailed behaviour as part of a large model; extract the detailed part and create a separate model for it.

Errors may result from incorrect instructions or parameters in the input data or from 'bugs' in the software. Erroneous results can only be minimised by scrutinising the instructions and input data for mistakes, and by thorough verification and calibration of the model. Verification can involve using the model to solve various standard groundwater flow problems and comparing the output with published solutions. Verification, and alteration of model and input data, continues until results agree with published solutions. Some examples of the application of numerical models to groundwater control schemes are given in Roberts *et al* (2009) and Bevan *et al* (2010).

## Calibration

Calibration is a trial and error procedure, which involves:

- running the model
- comparing the results with suitable available data (eg piezometer readings from the site investigation)
- adjusting the model parameters and boundary conditions within realistic ranges until there is an acceptable agreement between field observations and model output.

## Prediction

Once the model has been calibrated, it can be run to predict the results of interest (eg flow rate, rate of drawdown, distribution of drawdown). Parametric studies are often carried out to determine the possible range of predictions.

## Refinement

See also
3.4 Monitoring

The model can be refined by calibrating it against the results of monitoring (see Section 3.4). Refinement can be very useful when work is carried out in stages – the initial dewatering phases can be used to reduce uncertainty in the design of later works.

*Like any modelling method, the results of a numerical model can only produce results based on the conceptual model. If the conceptual model does not reflect actual conditions, the results are unlikely to be useful.*

With some numerical models, the size of the model grid and type of boundary may affect the calculated results. For example, the distance of influence (see Section 6.1.2) affects flow rate, but some numerical models will automatically calculate flow rate by assuming the distance of influence is exactly at the model boundary. This can give the strange result that different flow rates may be calculated from numerical models that are identical in every way except the distance from the wells to the boundaries. This may still be the case even if the boundaries of the model are much further away than the predicted distance of influence. The effect of varying the distance to the model boundaries, and of varying flow conditions at the boundaries, can be assessed by carrying out a parametric study (Box 6.1) (see Powrie *et al*, 1989, or Kofoed and Doran, 1995).

# 6.2 ESTIMATION OF STEADY-STATE FLOW RATE

See also
Figure 6.1 Design
Box 6.1 Sensitivity analysis
6.1.4 Numerical modelling
6.4 Time dependent analysis

Estimation of the total steady-state flow rate is an important step in any design. It can provide the basis for determination of the well capacity, number of wells, their depth, and the drawdown that will be achieved (see Figure 6.1). It is important to remember that the initial flow rate required to achieve the drawdown within the required time (see Section 6.4) may be much greater than the steady-state flow rate needed to maintain the drawdown in the long-term. This section concentrates on relatively simple analytical methods to estimate steady-state flow rate. In more complex situations, and where design parameters have been determined sufficiently accurately, flow rate could also be calculated by numerical methods (Section 6.1.4) or as part of a time-dependent analysis (Sections 6.4 and 6.5).

Flow rate calculations are normally carried out in SI units (ie permeability in m/s, and all dimensions, including distance of influence or radius of influence, in metres), which will produce flow rates in $m^3/s$. Conversion factors to other units are given in Datasheet 1 (Appendix A1). Derivations of many of the formulae in this section are given in Mansur and Kaufman (1962).

## 6.2.1 Equivalent well analysis

Groundwater control systems are generally installed either as rings around an excavation or in long lines alongside a trench. The equivalent well method models a ring of closely spaced dewatering wells as one large equivalent well and uses radial flow theory (Figure 6.5a and b). Long lines of closely spaced wells are modelled as equivalent slots using plane flow theory (Figure 6.5c). To carry out the analysis, the following points need to be identified:

1 Is the aquifer confined or unconfined?

2 Is the slot or well fully or partially penetrating?

3 Is flow to the system predominantly horizontal? Equivalent well methods are based on the assumption of horizontal flow. If an overlying or underlying source of recharge or a deep cut-off wall causes vertical flow, other methods may be more appropriate, such as flownets (Section 6.2.2) or numerical modelling (Section 6.1.4).

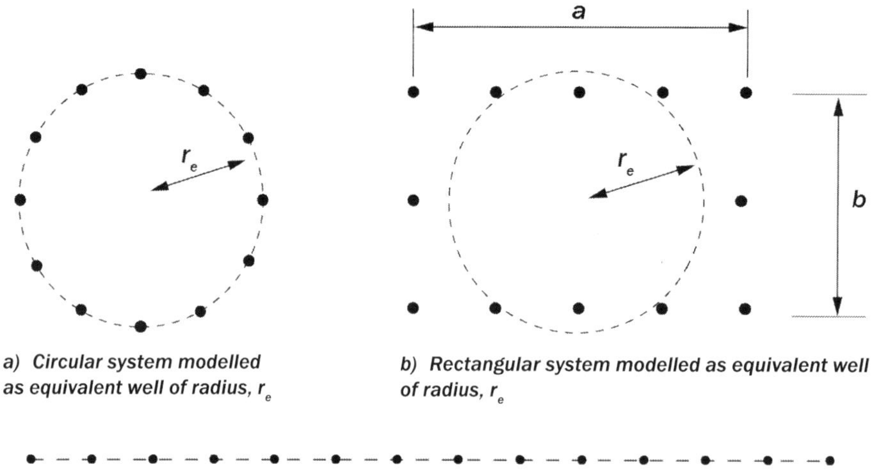


a) Circular system modelled as equivalent well of radius, $r_e$

b) Rectangular system modelled as equivalent well of radius, $r_e$

c) Long narrow system modelled as continuous slot between wells

*Figure 6.5 Equivalent wells and slots*

If the flow is predominately horizontal, Equations 6.6 to 6.8 presented here can be used.

For circular systems the radius of the equivalent well, $r_e$ , can be taken as the radius of the system. For rectangular systems of plan dimensions $a \times b$, $r_e$ can be estimated using Equation 6.6 (for further discussion of equivalent radius see Powrie and Preene, 1992).

$$r_e = (a + b)/\pi \tag{6.6}$$

The flow rate, $Q$, from a confined aquifer can be estimated using the Theim equation (Equation 6.7), and from an unconfined aquifer by using the Dupuit-Forcheimer equation (Equation 6.8).

## Fully penetrating wells

Confined conditions: $$Q = \frac{2\pi kD(H - h_w)}{\ln[R_o/r_e]} \tag{6.7}$$

Unconfined conditions: $$Q = \frac{\pi k(H^2 - h_w^2)}{\ln[R_o/r_e]} \tag{6.8}$$

where $k$ is the soil permeability, $D$ is the thickness of the confined aquifer, $H$ is the initial piezometric (or water table) level in the aquifer and $h_w$ is the piezometric (or water) level in the equivalent well (Figure 6.6). For the purposes of estimating flow rate to equivalent wells, $h_w$ should correspond to the target drawdown inside the excavation and not the drawdown inside individual wells, which may be affected by well losses (see Section 6.3.1).

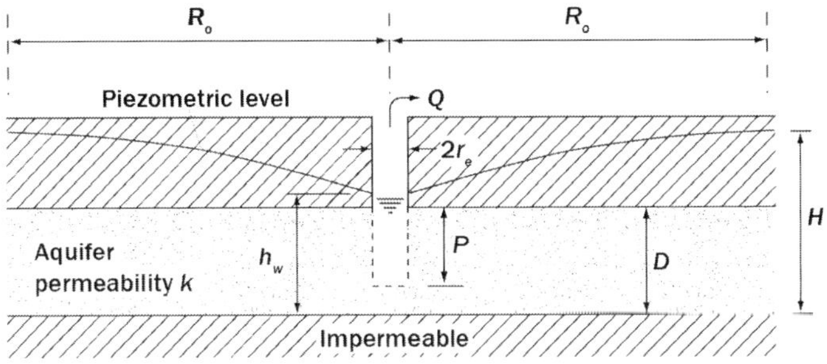


*a) Confined aquifer*

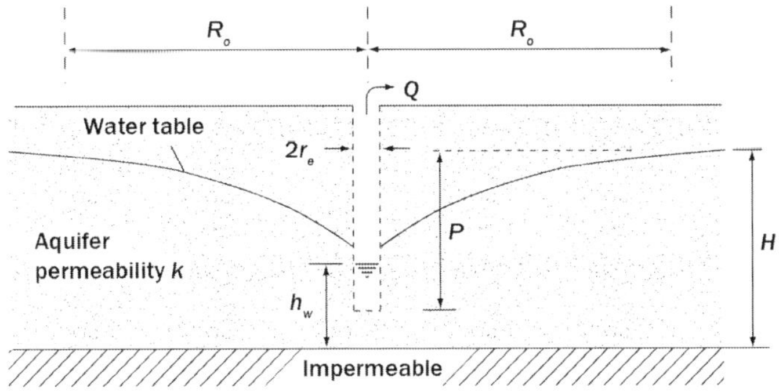


*b) Unconfined aquifer*

***Figure 6.6*** ***Idealised radial flow to wells***

$R_o$ is the radius of influence and for the purposes of flow rate calculations can be estimated from the empirical formula of Sichardt (Powers *et al*, 2007):

$$\text{Radial flow:} \quad R_o = C(H - h_w)\sqrt{(k)} \tag{6.9}$$

where C is an empirical calibration factor and $(H - h_w)$ is the drawdown in the equivalent well (ie the target drawdown in the excavation). If $(H - h_w)$ and $k$ are in metres and m/s respectively, to obtain $R_o$ in metres, C is usually taken as 3000. Because $R_o$ appears within a logarithmic term in the denominator of Equations 6.7 and 6.8, calculated flow rates are not excessively sensitive to different values of $R_o$. If the range of values of $R_o$ is very wide, conservative values (ie at the low end of the possible range, which will predict larger flow rates) can be used; alternatively a sensitivity analysis could be carried out (see Box 6.1).

## Partly penetrating wells

A partially penetrating well is defined as one that penetrates depth $P$ below the top of a confined aquifer or below the original water table of an unconfined aquifer. The flow rate $Q_{pp}$ from such a well will be less than the flow rate $Q_{fp}$ from a fully penetrating well:

$$Q_{pp} = BQ_{fp} \tag{6.10}$$

where $B$ is a partial penetration factor with values between 1 (for a fully penetrating well) and zero (when $P = 0$). For individual wells with a radius, $r_w$, of less than approximately 0.5 m, Figure 6.7 shows partial penetration factors, $B$, developed by Kozeny for confined aquifers and by Borelli for unconfined conditions (Mansur and Kaufman, 1962). Figure 6.7 assumes isotropic permeability conditions ($k_v = k_h$). In a deep or thick aquifer and for anisotropic conditions where $k_v << k_h$, the influence of partial penetration on the yield of a well is likely to be significantly diminished compared to these graphs.

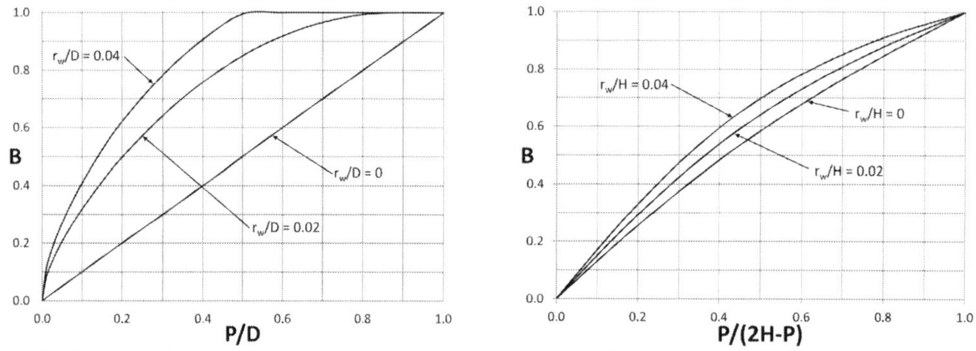


*a) Confined aquifer* *b) Unconfined aquifer*

*Figure 6.7 Partial penetration factors for wells (after Mansur and Kaufman, 1962)*

These factors cannot be directly applied to equivalent wells of large diameter, but they do illustrate the relationship between flow rate and penetration. If partial penetration is a major factor on a site (eg when dewatering in very deep aquifers), numerical modelling (Section 6.1.4) should be considered.

Long lines of closely spaced wells (or two parallel lines of wells) with flow from each side can be modelled as slots using the plane flow formulae based on the work of Chapman (1959) (Figure 6.8).

## Fully penetrating slots

Confined conditions: $$Q = \frac{2kDx(H - h_w)}{L_o} \qquad (6.11)$$

Unconfined conditions: $$Q = \frac{kx(H^2 - h_w^2)}{L_o} \qquad (6.12)$$

where $L_o$ is the distance of influence, $x$ is the linear length of slot, and all other terms are as defined previously. Because $L_o$ appears in a linear term in the denominator, Equations 6.11 and 6.12 are sensitive to $L_o$; doubling $L_o$ will halve the flow rate. Mansur and Kaufman (1962) state that Sichardt's formula (Equation 6.9) can be used to obtain $L_o$, taking $C$ as between 1500 and 2000. However, in high permeability soils where very large values of $L_o$ are calculated, caution is needed. Chapman's equations (6.10 and 6.11) were developed for ratios of $L_o/H$ of less than 5 and may not be suitable for application where $L_o$ is very large; flow rates may be significantly underestimated. A sensitivity analysis (Box 6.1) could be used to demonstrate these uncertainties.

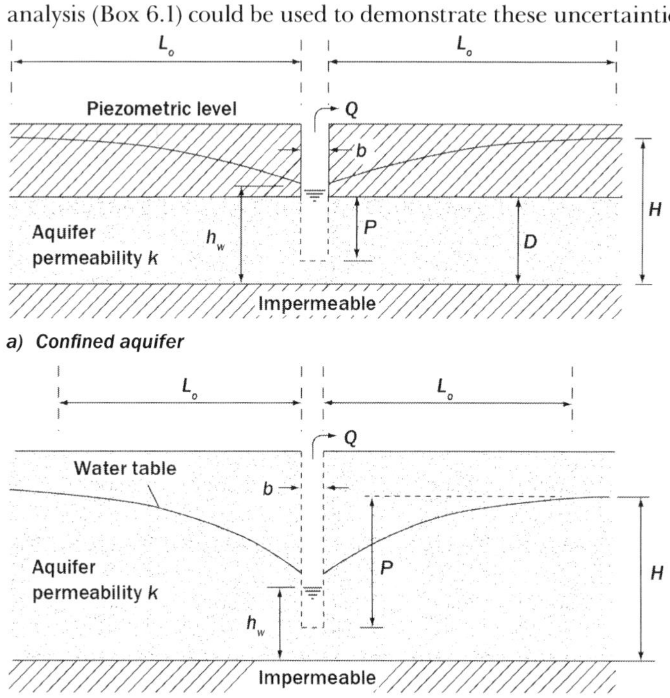


*a) Confined aquifer*

*b) Unconfined aquifer*

*Figure 6.8 Idealised plane flow to slots*

## Partially penetrating slots

Confined conditions: $$Q_{pp} = \left[0.73 + 0.27\frac{P}{H}\right]\frac{kx(H^2 - h_w^2)}{L_o} \quad (6.12)$$

Unconfined conditions: $$Q_{pp} = \frac{2kDx(H - h_w)}{L_o + \lambda D} \quad (6.13)$$

where $Q_{pp}$ is the flow rate from a partially penetrating slot, and all terms are as defined previously apart from $P$, which is the depth of penetration of the slot below the top of a confined aquifer or below the original water table in an unconfined aquifer, and $\lambda$ is the partial penetration factor for confined slots, which can be obtained from Figure 6.9.

The above equations assume the slot is of infinite length and flow is purely plane to the sides of the slot. This may be a reasonable assumption for lines of wells alongside a long trench, but may be less appropriate for excavations of shorter length; plane flow may occur at the sides, but radial flow is likely at the ends (Figure 6.10).

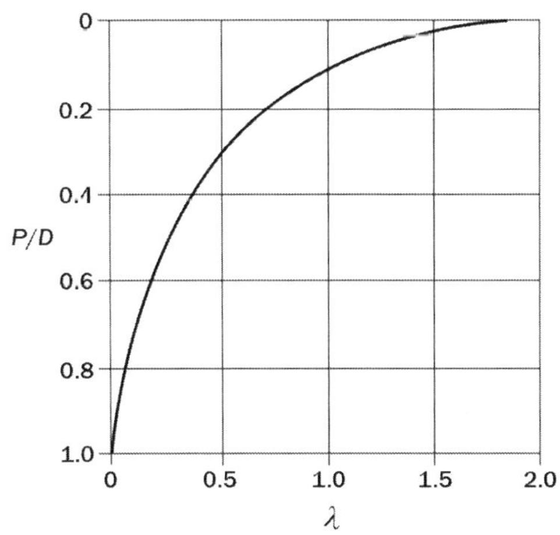


*Figure 6.9 Partial penetration factors for confined flow to slots (after Mansur and Kaufman, 1962)*

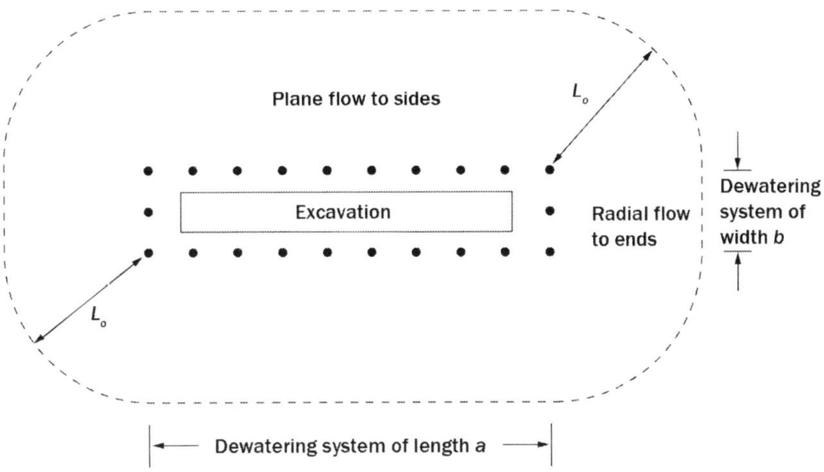


*Figure 6.10 Plane and radial flow to excavations*

It may be appropriate to calculate flow rate to the sides (where flow is plane) as if it were a slot of length $a$, and add to that the radial flow to the ends by assuming the ends act as a well of radius $b/2$. For confined aquifers, a shape factor $G$ can be used to calculate flow rate from rectangular equivalent wells of plan dimensions $a \times b$ using Equation 6.14, where $(H - h_w)$ is the drawdown inside the excavation. Flow to rectangular excavations is discussed further by Powrie and Preene (1992). Values of $G$ for different excavation aspect ratios $a/b$ and distance of influence $L_o$ are given in Figure 6.11.

## Plane and radial flow to rectangular systems

Confined conditions: $$Q = kD(H - h_w)G \quad (6.14)$$

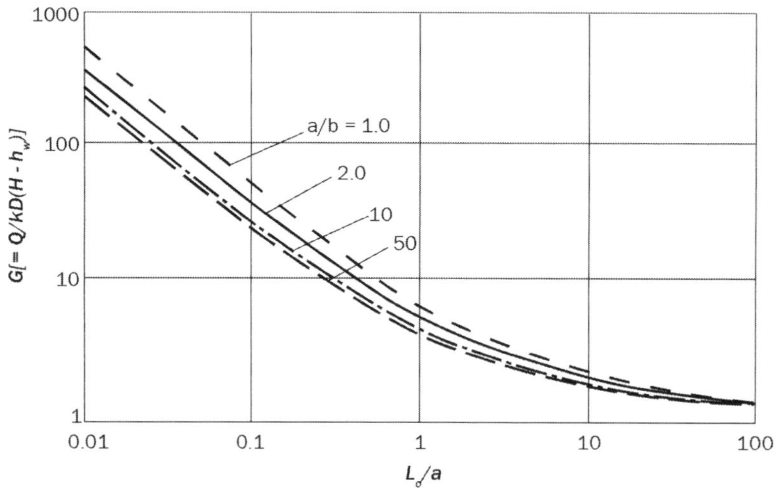


*Figure 6.11 Shape factor for confined flow to rectangular equivalent wells (from Powrie and Preene, 1992)*

## 6.2.2 Flownets

**See also**
Box 6.7 Flowpath graphic

The equivalent well method is suited to situations where groundwater flows are predominantly horizontal, and aquifer boundary conditions are not complex. For irregular geometries or complex boundary conditions, flownet analysis can be useful. A flownet is a graphical solution to the mathematical equations controlling two-dimensional groundwater flow. Many numerical models (Section 6.1.4) will produce graphical output (see Box 6.7) in the form of flownets, but this Section is principally concerned with trial and error solutions by hand sketching. Flownet sketching is described by Cedergren (1989).

In general, a flownet consists of a series of flowlines (idealising the path of water particles) and, perpendicular to these, a series of equipotential lines (showing the distribution of groundwater head). Because they produce graphical representations of the distribution of groundwater heads and flow patterns, flownets can be useful for judgmental adjustments of design (eg moving well locations to intercept localised groundwater flows). A flownet can be used to solve either vertical problems (eg seepage beneath cofferdams) or horizontal problems (eg recharge from a river), but it is essentially a two-dimensional solution for plane flow only. The method is unsuitable for analysis in three dimensions. In practice, flownets are often used as a preliminary method for rough calculations to assess the effect of possible aquifer boundary conditions. A case history describing the use of flownets is given by Powrie and Roberts (1995). For soils of anisotropic permeability, flownets can be drawn to a transformed scale to model flow patterns (Cedergren, 1989).

## 6.2.3 Seepage into cofferdams

**See also**
1.2.6 Instability
6.1.4 Numerical modelling

If a shallow stratum is more permeable than the underlying soil, a cut-off wall may be used to form a cofferdam to exclude shallow groundwater (Figure 6.12). The cofferdam forces groundwater to take a longer path into the excavation, thereby reducing the flow rate and decreasing the seepage gradients (which can cause instability, see Section 1.2.6). If the overlying layer is at least two to three orders of magnitude more permeable than the underlying stratum, negligible drawdown occurs in the upper layer, which can be considered as a constant head boundary. The overlying layer could be a coarse gravel ($k \approx 10^{-3}$ m/s) over silty sand ($k \approx 10^{-5}$ m/s), or it may be open water (theoretically of infinite permeability) overlying a beach deposit, eg where a pipeline is being built across a foreshore.

Because there is significant vertical flow, equivalent well methods should not be used. Flow rate could be estimated by flownet sketching or using a numerical model (Section 6.1.4) with appropriate boundary conditions. For long cofferdams, where flow is plane to the sides, standard solutions have been developed for flow rate *Q* (Figure 6.12). These solutions are intended to estimate flow rate for a cofferdam of given penetration. The cofferdam has to be designed to avoid the risk of piping or seepage failure at the base, particularly in the corners, see Williams and Waite (1993).

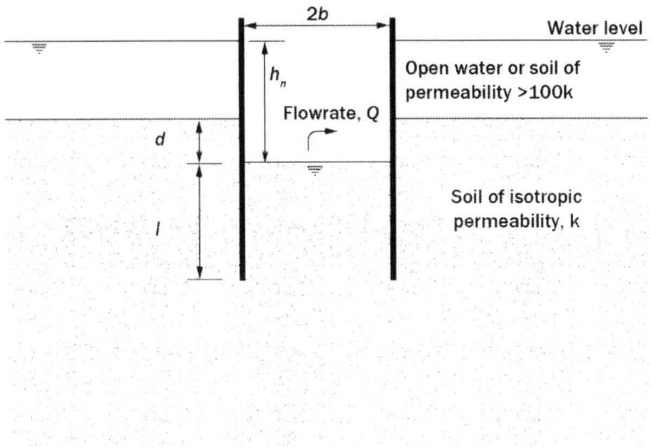


*Figure 6.12 Geometry for plane seepage into a long cofferdam (from Carter, 1982)*

Flow rate is calculated by determining the seepage factor, *m*, from Figure 6.13a and obtaining *Q* from Figure 6.13b. The calculated flow rate is per metre run of cofferdam. For very long cofferdams (where end effects can be neglected) the total flow rate is determined by multiplying the result from Figure 6.13b by the length of the cofferdam. For shorter cofferdams end effects should be considered. Powrie and Preene (1992) showed that if the cofferdam is more than 10 times longer than it is wide, the total flow rate can be calculated by assuming plane flow to the sides only (ie end effects can be neglected). For shorter cofferdams, where the length is between one and five times the width, total flow rate can be calculated by assuming plane flow to all four sides of the excavation (ie corner effects can be neglected).

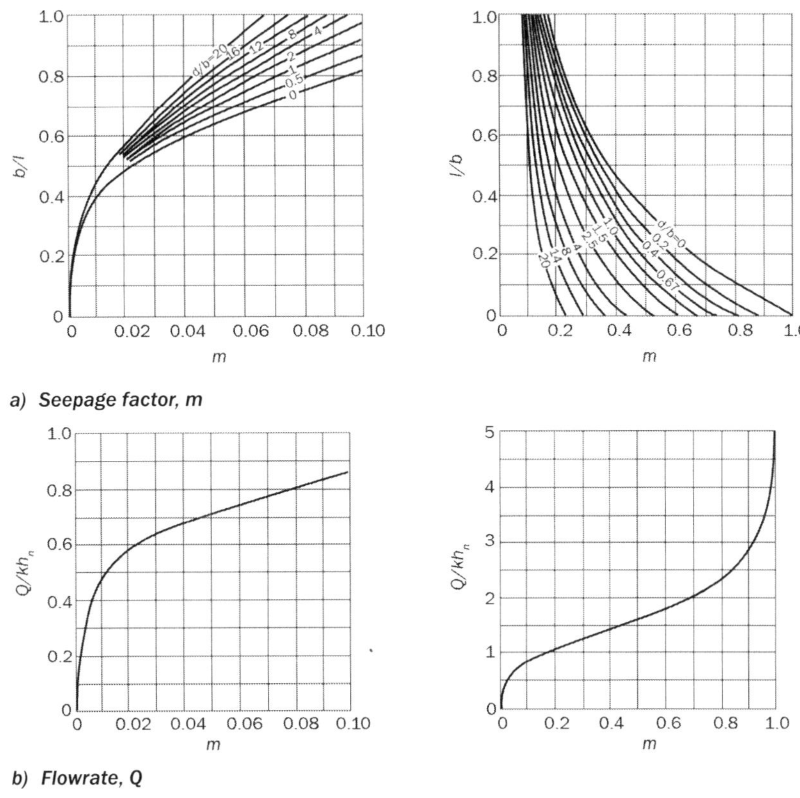


*Figure 6.13 Relationship between discharge and geometry for plane seepage into a long cofferdam (from Carter, 1983)*

## 6.2.4 Other methods

See also
6.1.4 Numerical modelling
6.5.2 Drawdown patterns
7.2 Observational method

Analogue methods use physical models to analyse groundwater flows. The flow of electricity is governed by the same mathematical equations as groundwater flow, so electrical resistance networks can act as analogue models of flow in aquifers (Rushton and Redshaw, 1979). The rapid advances in the power and convenience of computer and numerical methods (Section 6.1.4) has meant that electrical analogue methods are rarely used today. A more recent application is described by Knight *et al* (1996).

The method of superposition, which estimates the distribution of drawdown around a group of wells, can also be used to calculate the flow rate required from each well to achieve the target drawdown at chosen locations (see Section 6.5.2).

On sites where ground conditions may be very complex, or if a full-scale site investigation is precluded by time or practical limitations, the Observational Method (Nicholson *et al*, 1999) is sometimes used. This involves estimating the dewatering system that will probably be required, installing that system and then using the results of field monitoring (see Section 3.4) to modify the system until the target performance is achieved (Section 7.2).

# 6.3 DESIGN OF WELLS AND FILTERS

## 6.3.1 Flow of groundwater to wells

Section 6.2 describes methods for determining the total extraction flow rate from a groundwater control system necessary to achieve the desired drawdown. To complete the design of the system, the number of the wells must be determined and the following parameters specified:

- the depth of the wells
- the well screen and filter pack
- the pump capacity required (determined from the estimated well yield)
- the well liner necessary to accommodate the chosen pump.

Although the flow lines to a line of wells are parallel remote from the wells, close to each well the flow is radial. As the groundwater approaches the well, the radial flow leads to a reduction in the available area of flow and a corresponding increase in the pore water velocity. This problem is exacerbated in shallow unconfined aquifers because of the drawdown at the well, Figure 6.14. The increase in pore water velocity results in well losses. Well loss is the difference between the water level inside the well and that in the soil immediately outside (Figure 6.14). Well losses may be large; in soils of moderate to low permeability it is not unusual for the drawdown outside a well to be less than half of that inside the well.

Well losses can be minimised by adequate well development (Section 2.1.5), by specifying wells with sufficient wetted screen area, and by increasing the permeability of the ground close to the well screens by providing an appropriate granular filter.

## 6.3.2 Well depth

See also
6.5.2 Drawdown patterns

The principal factors that determine well depth are summarised in Box 6.8. These are critical, as the required well depth has considerable influence on the selection of the pumping method. Selection of well depth can be complex, but in practice, the following rules of thumb are often used:

- Wellpoint systems are generally installed at close spacing, 1 m to 3 m, to penetrate to between 1 and 3 m below the proposed excavation level.
- Deepwells are usually installed at rather wider spacing, 10 m to 30 m, to penetrate to

approximately twice the required drawdown level (ie if the groundwater level is close to ground level, deepwells would generally penetrate to twice the depth of excavation).

- Ejector well spacing and depth generally falls between that of wellpoint and deepwell systems.

The wells have to be deep enough to generate the required drawdown after allowing for well losses and the accommodation of pumping equipment. The drawdown just outside the well should be such that, when combined with the drawdown produced by other wells in the system, the design drawdown pattern is achieved (see Section 6.5). In shallow or thin aquifers the effective well depth may be limited by an impermeable layer at the base of the aquifer. In this case, the design may have to be modified so that the system comprises a greater number of shallower wells, possibly of reduced capacity.

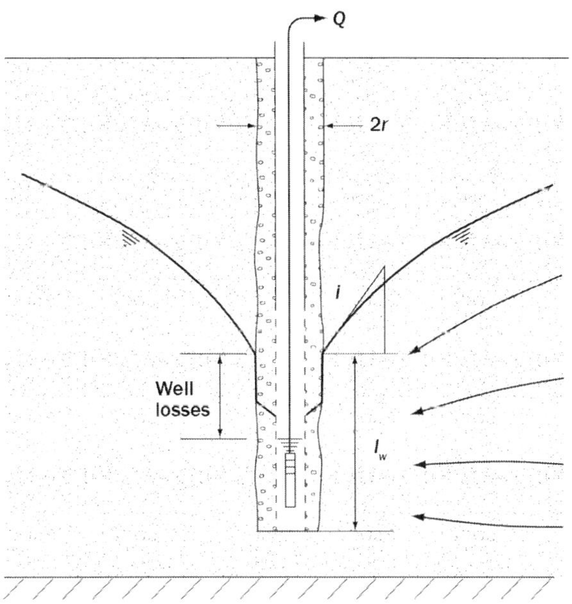


*Figure 6.14 Reduction of area of flow and well losses as groundwater approaches a well*

*Box 6.8* *Principal factors affecting selection of well depth*

1 ***Required drawdown***

Wells should be deep enough to generate the required drawdown in the area of the excavation.

2 ***Screen area***

Wells should be deep enough so that sufficient wetted area of screen is available to achieve the intended well yield (Section 6.3.4).

3 ***Confined aquifers***

Wells should be deep enough to penetrate into any confined aquifers requiring pressure relief.

4 ***Partial penetration***

It may not be necessary for wells to penetrate to the full depth of a deep aquifer; sometimes partially penetrating wells can achieve the same drawdowns at lower discharge flow rates.

5 ***Control of overbleed***

'Toeing-in' wells into an impermeable interface can minimise the overbleed flow (see Figure 2.4). Deepwells can be installed so that the pump intake is below the top of the impermeable interface (see Case study 7.5), although there is a risk of the pump overheating.

6 ***Stratified soils***

Dewatering wells should penetrate to the most permeable stratum, unless this is very deep. Drainage of a permeable layer can lead to underdrainage of overlying finer soils using relatively few wells (see Case study 7.1).

## 6.3.3 Design of filters

**See also**
5.3.5 Particle size analysis

Rules for specifying granular filters, principally for use in dams, were first put forward by Terzaghi and Peck in 1948 (Terzaghi *et al*, 1996) and have been exhaustively studied since then. Later studies by Sherard *et al* (1984a, 1984b) and Kenney *et al* (1985) support Terzaghi's filter rules. The criteria for specifying filters are based on the PSD curves (see Box 5.5) of samples of the material to be filtered, in this case the aquifer surrounding the well. However, caution should be used in case finer particles have been washed out of the sample (see Section 5.3.5).

An effective granular filter should be:

1 Fine enough to prevent persistent movement of fines from the aquifer. Experiments have shown that aquifer particles smaller than 0.1 × $D_{15filter}$ can pass through a filter in normal service conditions.

2 Sufficiently coarse so that it is significantly more permeable than the aquifer, in order to minimise head losses close to the well. Permeability of granular material is related to the size of the fine fraction on a PSD grading curve (see Section 5.3.5).

3 Sufficiently uniform to allow installation with minimum risk of segregation (criteria for well filters are more strict than for dams because annular well filters are generally thinner and placement is more difficult to control).

Criteria for granular filters for sands are given in Box 6.9. Strict application of the rules in Box 6.9 to the most fine-grained of the PSD curves for the aquifer can result in wells which are difficult to develop and may give poor yields. The most effective granular filter is one which is as coarse as possible, and therefore as permeable as possible, but is just able to prevent persistent movement of fines from the aquifer.

*Box 6.9 Criteria for granular filters for sands*

- **Pore size of the filter should be sufficiently small to prevent persistent movement of fines from the aquifer**

  $D_{15filter} \leq 5 \times D_{85aquifer}$ (this is known as Terzaghi's filter criterion)

  Ideally this should apply to the coarser side of the filter grading envelope and the finer side of the aquifer grading envelope.

- **The filter should be significantly more permeable than the aquifer.**

  $D_{15filter} > 4 \times D_{15aquifer}$

  Ideally this should apply to the fine side of the filter grading envelope and the coarse side of the aquifer grading envelope. In addition the filter should not contain more than five per cent fines (<63 μm).

- **The filter should be sufficiently uniform to minimise risk of segregation during placement.** Powers *et al* (2007) indicates that with care filters can be successfully placed by gravity, providing the uniformity coefficient $U$ ($U = D_{60}/D_{10}$) is such that:

  $U_{filter} < 3$

  This can be relaxed provided $U_{filter} < U_{aquifer}$. If $U_{filter} > 3$, filter placement by tremie pipe is recommended.

- **The well slot size should be sufficiently small to retain the filter.**

  Well slot size $\approx D_{10filter}$

**Notes**

1 These criteria are applicable to filters for sands, and sandy silts with $D_{85aquifer} \geq 0.1$ mm (Sherard *et al*, 1984b). The criteria can be relaxed for finer soils – see comments on filters for silts.

2 For sands with $D_{40aquifer} > 0.5$ mm and $U_{aquifer} > 3$, it may be preferable to form a natural filter pack by well development (Misstear *et al*, 2006) – see comments on natural filters.

3 Movement of fines is minimised by screening for the finest stratum present, but in layered aquifers this may limit the capacity of the well – see comments on filters for layered soils.

4 Filters should not be gap-graded (ie having one or more near horizontal sections of the PSD curve).

5 If the aquifer is gap-graded, the filter should be specified using the PSD of the fine fraction of the aquifer material.

6 Some published filter criteria place a limit on the $D_{50aquifer}/D_{50filter}$ ratio. These are not founded on a sound theoretical or experimental basis (Sherard *et al*, 1984a).

7 It is not necessary that the PSD curve of the filter be similar in shape to the PSD curve of the aquifer as some published filter criteria suggest (Sherard *et al*, 1984a).

Granular filters have been shown to be effective even when as thin as 12 mm (Sterrett, 2009). For dewatering wells, however, annular thicknesses of less than 50 mm are rarely used. It is good practice to fit centralisers to the screen so the annulus is equally spaced around the well screen over the full length. The filter should extend above the level of the screen to allow for some loss or compaction of the filter material during placing and well development. A filter thickness greater than 100 mm to 150 mm is not recommended because this can make it difficult to develop the wall of the borehole.

Well screens can be obtained with a specified granular filter material bonded direct to the screen; these resin-bonded screens can prove useful when a granular filter is critical, but the annulus is too small to allow conventional methods of placement.

## Natural filter packs

See also
2.1.5 Well development

Coarse, widely graded soils, such as gravelly sands and sandy gravels, can be self-filtering, so it is sometimes not necessary to introduce an artificial filter pack. In these soils an effective natural filter pack can be produced by selecting an appropriate slot size combined with development (see Section 2.1.5) to remove the fine fraction of the soil close to the well. Soils can be considered to be natural filters provided $D_{40aquifer} > 0.5$ mm and $U_{aquifer} > 3$ (Misstear *et al*, 2006). The optimum slot size for the development of natural filter packs is the maximum, which does not lead to continuous pumping of fines. Published recommendations are for slot sizes in the range $D_{30aquifer}$ to $D_{70aquifer}$. For most applications a slot size of $D_{40aquifer}$ to $D_{50aquifer}$ is acceptable, but in widely graded soils, if the maximum well yield per unit length of screen is required, a slot size in the range $D_{60aquifer}$ to $D_{70aquifer}$ could be considered.

## Filters for silts

See also
2.2 Pore water pressure control systems

When carrying out vacuum-assisted drainage of fine-grained soils (Section 2.2), it is sometimes necessary to specify filters suitable for silts. Application of the criteria in Box 6.8 can lead to finer filter materials and screen sizes than those necessary in practice. Sherard *et al* (1984b) suggest the following criteria:

- for silts and clays with some sand content, $D_{85aquifer} > 0.1$ mm, the criteria in Box 6.9 apply (implying $D_{15filter} \leq 0.5$ mm)
- for fine silts without significant sand content, low plasticity and $D_{85aquifer}$ of 0.03 mm to 0.10 mm, sand filters with average $D_{15filter} \leq 0.3$ mm are conservative
- silts with $D_{85aquifer} < 0.02$ mm are not common in nature. For these soils a filter with average $D_{15filter} \leq 0.2$ mm is conservative.

These criteria were developed for critical filters for cores in earth dams. It is possible that they may be slightly conservative for well filters. This can only be confirmed by practice and experience in the conditions under consideration.

## Filters for layered soils

PSD curves for samples from a particular aquifer formation will often show considerable variation. A grading envelope can be prepared for a particular formation by plotting all of the PSD curves together. Also, a well may penetrate several strata with different grading envelopes. In order to minimise the risk of continuous pumping of fines, the filter design should be based on the finest grading, but this may limit the capacity of the wells. The use of filters and screens, which vary with depth through the differing strata is good in principle, but is difficult to achieve in practice. It may be more effective to use plain well liner through the finer strata and to base the screen and filter design on the coarser soils. No general criteria are available and experience and judgement are required when putting theory into practice.

## Geotextile filters

Well screens wrapped in geotextile are sometimes used in dewatering applications. Appropriately constructed layered geotextile screens can provide a small size of opening combined with the benefit of a large screen open area (see Table 2.8). Woven filter fabrics have a measurable opening size (equivalent to the slot size on a conventional screen) and the smaller openings can allow the use of natural filter packs in much finer soils than is possible using conventional slotted screens. Wells can be more efficient as a result and both cheaper and easier to install. Granular filters can be used in conjunction with geotextile filters (the criteria in Box 6.9 apply). Guidance may also be available from manufacturers. Non-woven filter fabrics are not generally considered appropriate for well screens because difficulties may arise in development. The filter properties of geotextiles are considered in detail by Hausmann (1990) and Kennedy *et al* (1988).

## Formation stabilisers

Dewatering wells installed in weak rock, such as soft or weak chalk or poorly cemented sandstone, sometimes require a granular formation stabiliser in the borehole annulus around the well screen. The principal flow from such strata is generally from fissures and a coarse screen is necessary to minimise head losses where a fissure is intercepted. The purpose of the formation stabiliser is to fill the annulus to prevent the strata collapsing on to the screen. In poorly cemented sandstone the stabiliser may also be necessary to minimise loss of sand from the strata. No general criteria are available and local experience plays a large part in the selection and specification of formation stabilisers.

# 6.3.4 Estimation of individual well yields

**See also**
Figure 6.14 Flow to a well

A seepage analysis (Section 6.2) is generally used to predict the total expected extraction flow for a groundwater control scheme. Flow rates to individual wells need to be assessed to estimate the number of dewatering wells required. Prediction of individual well yields is an inexact science, which has received relatively little research attention. The following factors affect the yield of a well:

- hydraulic characteristics of the aquifer, eg permeability
- wetted length of well screen
- effective radius of well
- screen and filter specification
- correct well development (see Box 2.3).

Assuming that the screen and filter have been selected to optimise yields using the procedures set out in Section 6.3.3, Darcy's law can be applied to the boundary of an individual well filter to give:

$$q = 2\pi r l_w k i \qquad (6.15)$$

where $q$ is the individual well yield, $r$ is the effective radius of the well (usually taken as the drilled borehole radius), $l_w$ is the wetted length of well screen, $k$ is the aquifer permeability and $i$ is the hydraulic gradient at entry to the well (see Figure 6.16). Several authors, such as Powers *et al* (2007) and Hausmann (1990), quote Sichardt's formula for estimating the maximum hydraulic gradient at entry into the well:

$$i_{\max} = \frac{1}{15\sqrt{k}} \qquad (6.16)$$

When applied in Equation 6.15 this formula gives a reasonable first estimate for the yield from wells in aquifers with a permeability above about $10^{-4}$ m/s. Application of this formula gives similar results to the limiting screen entrance velocity approach advocated in the water supply industry for well design in aquifers of permeability $2.3 \times 10^{-4}$ to $2.8 \times 10^{-3}$ m/s (Howsam *et al*, 1995, Appendix 2). For high permeability aquifers, $k > 10^{-3}$ m/s, the potential well yields may be so large that flow rates are controlled by the capacity of the pump rather than the well.

For aquifer permeabilities below about $10^{-4}$ m/s Equation 6.16 appears to give unrealistically high values of hydraulic gradient (and hence well yields). Preene and Powrie (1993) analysed data from a number of case studies where vacuum pore water pressure control systems were used in soils of permeability $10^{-6}$ to $5 \times 10^{-5}$ m/s, and found that measured individual well yields can vary by a factor of more than 100 at a given site. The method of installation was found to be important, with jetted wells giving better performance than those installed by rotary or cable percussion drilling. Despite the wide variation, some consistency was found by considering average hydraulic gradients and it was shown that $i_{max}$ was approximately 10 for sealed ejector wells and 4 for vacuum wellpoints.

These results have been combined to produce Figure 6.15, which shows the relationship between aquifer permeability and well yield per unit length of wetted screen per unit effective well radius. Figure 6.15 can be used to provide a first estimate of average individual well yields, but should not be relied upon until supported by appropriate practical experience.

After determining the maximum well yield for the assumed borehole drilled radius, *r*, or estimating it from pumping tests, the proposed well diameter should be checked. The well screen diameter that can be used in the borehole and still allow room for an adequate filter (see Section 6.3.3) should be determined. The pump manufacturer's specification (or Table 2.6) will show if a pump of adequate capacity can be installed, operated and removed in that diameter of well screen. If the required pump is too large to fit down the screen, a larger drilling diameter will have to be specified and the above checks repeated until a satisfactory result is obtained.

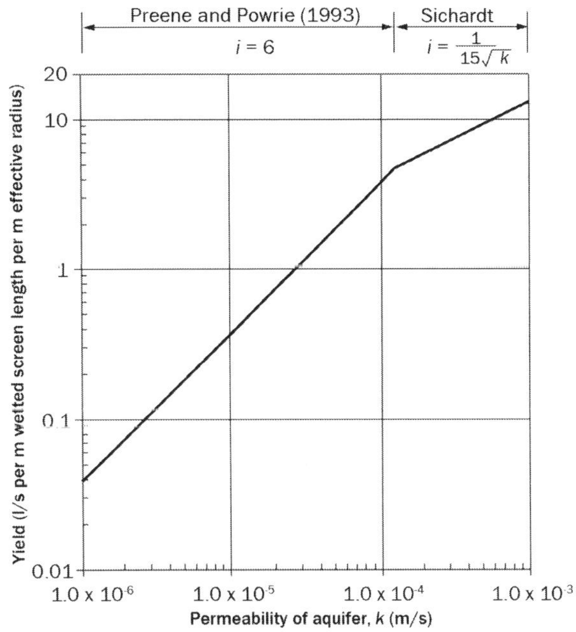


***Figure 6.15 Approximate maximum well yields***

# 6.4 ESTIMATION OF TIME–DRAWDOWN RELATIONSHIP

## 6.4.1 Information required for design

**See also**
1.2.6 Objectives of groundwater control
6.1.4 Numerical modelling
6.5.2 Drawdown patterns

This section addresses the time taken for the groundwater control system to achieve the required drawdown, which in some cases has an important influence on the timing or sequencing of a construction process.

The methods of analysis described in Section 6.2 are concerned with steady-state conditions, ie when the drawdown and flow rate do not change with time. The analyses described in this section imply that in an ideal infinite aquifer a true steady-state will never be achieved. In practice, however, rainfall and other sources of recharge mean that a stage will usually be reached when continuing increases in drawdown with time are almost imperceptible. The methods described in Section 6.5.2 may also be used in soils of moderate to high permeability to estimate the time-dependent drawdown pattern around a group of wells.

In unconfined aquifers consisting of coarse-grained soils, a lowering of the groundwater level is accompanied by the drainage of a potentially substantial volume of pore water from the soil and its replacement by air. Owing to capillary effects, unconfined aquifers consisting of fine-grained soils tend to remain saturated unless the drawdown of the water table is large, and are therefore depressurised rather than dewatered. In confined aquifers, pumping groundwater from wells reduces the pore water pressure but does not actually dewater the soil, which remains saturated. If the soil remains saturated, any drainage of pore water is a result of a change in pore volume as the soil consolidates.

This distinction between the behaviour of coarse-grained unconfined aquifers on the one hand, and fine-grained unconfined aquifers and confined aquifers on the other, is discussed in Section 1.2.6. The two mechanisms – desaturation (dewatering) and consolidation (depressurisation) – are fundamentally different, and are analysed in different ways.

See also

Table 6.1 Conceptual model

Table 6.4 Estimation of soil stiffness

In addition to the data on ground and groundwater conditions, soil permeability, sources of recharge and excavation geometry summarised in Table 6.1, the following information is needed to estimate the time–drawdown relationship:

- for unconfined coarse-grained aquifers, the storage coefficient, $S$ (Section 6.1.2), indicates the volume of water that will drain by gravity from the soil pores, per m$^3$ of soil dewatered
- for unconfined fine-grained aquifers and all confined aquifers, the permeability, $k$, and the stiffness of the soil in one-dimensional compression, $E'_o$ (see Section 6.6.2 and Table 6.4).

In fine-grained soils, the permeability and the stiffness in one-dimensional compression may be combined to give the consolidation coefficient, $c_v = kE'_o/\gamma_w$, where $\gamma_w$ is the unit weight of water and $c_v$ is approximately related to $S$ by $c_v = kD/S$, where $D$ is the aquifer thickness. Coarse-grained soils are comparatively stiff and permeable: volume changes from consolidation are therefore generally small and occur very quickly – indeed, such soils are not generally thought of as consolidating.

See also

6.3.1 Well losses

The analyses presented in this section assume that there are no well losses (see Section 6.3.1), ie the water level inside the well is the same as in the soil immediately outside the well. Losses may be significant for individual wells, but for the methods presented in this section, where equivalent wells or slots are used to model groups or lines of wells, losses are likely to be sufficiently small not to affect the accuracy of the drawdown calculations significantly (Powrie and Preene, 1994a).

This section concentrates on the relatively simple analytical methods that can be used to estimate the time taken to achieve the required drawdown as the soil either desaturates or consolidates by drainage of water from the pores. In more complex situations numerical methods may be used (Section 6.1.4), provided that:

- the program is capable of modelling the relevant soil behaviour (ie desaturation or consolidation)
- the soil parameters and boundary conditions have been determined sufficiently accurately.

## 6.4.2 Rate of drawdown in low permeability soil

When a pumped well system is installed directly into a low permeability soil, the drawdown gradually extends laterally from the line or ring of wells at a rate governed by the consolidation characteristics of the soil. Standard consolidation solutions may be used to estimate the progress of the drawdown curve with time (Powrie and Preene, 1994a). Each successive curve represents a graph of drawdown $s$ against distance from the line of wells at a time $t$ after the start of pumping, and is known as an isochrone. Isochrones can be presented in dimensionless form, using a dimensionless time factor, $T$, related to the consolidation characteristics of the soil (see Powrie, 2013).

### Plane flow

For long excavations in which water is removed from the soil by horizontal plane flow to a line of closely spaced wells, idealised as an equivalent pumped slot (eg Box 6.4), the non-dimensionalised solution to the consolidation problem may be represented by a single parabola, as shown in Figure 6.16.

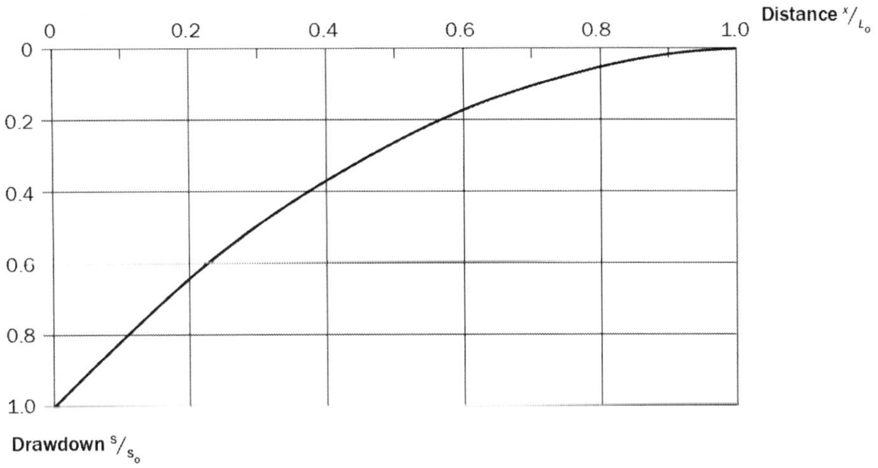


*Figure 6.16 Dimensionless drawdown curve for horizontal plane flow to a line of wells acting as a pumped slot in a low permeability soil*

Figure 6.16 shows the normalised drawdown, $s/s_o$, against the normalised distance from the line of wells, $x/L_o$, where $s$ is the drawdown at distance $x$, $s_o$ is the drawdown in the soil immediately outside the slot and $L_o$ is the current distance of influence. $L_o$ varies with time according to Equation 6.17 (Powrie and Preene, 1994a):

$$L_o = \sqrt{12c_{hv}t} \tag{6.17}$$

where $c_{hv}$ is the consolidation coefficient for vertical compression under horizontal drainage flow (and $c_{hv} = k_h E'_o / g_w$), and, in addition to the terms already defined, $k_h$ is the horizontal permeability.

The drawdown $s$ at a distance $x$ from the pumped slot at a time $t$ after the start of pumping can be estimated by determining $L_o$ at time $t$ from Equation 6.17 and then using Figure 6.18. Figure 6.18 has been calculated assuming:

- a line of wells close enough to act as a single equivalent pumped slot
- a uniform soil stratum with constant soil parameters $k_h$, $E'_o$ and $c_{hv}$
- purely horizontal flow
- no sources of vertical or horizontal recharge within the current distance of influence, $L_o$, of the line of wells
- no well losses (seepage face effects)
- a drawdown curve (isochrone), which is parabolic in shape (this is reasonable for plane flow).

## Radial flow

For a dewatering system idealised as an equivalent pumped well of radius, $r_e$, the numerical solution obtained by Rao (1973) may be used to develop isochrones of normalised drawdown, $s/s_o$, against the normalised distance from the centre of the equivalent well, $r/r_e$ (Figure 6.19), where $s$ is the drawdown at a radius $r$ and $s_o$ is the drawdown imposed in the soil immediately outside the equivalent well (ie at radius $r_e$). Here, the isochrone is plotted for different values of the dimensionless radial time factor, $T_r$:

$$T_r = \frac{c_{hv}t}{r_e^2} \tag{6.18}$$

where $r_e$ is the radius of the equivalent well, $t$ is the elapsed time, $c_{hv}$ is (as in the case of plane flow, above) the consolidation coefficient for vertical compression with horizontal drainage flow, and all other terms are as already defined.

The drawdown at a distance $r$ from the centre of an equivalent well of radius $r_e$ at a time $t$ after the start of pumping can be estimated by determining the time factor $T_r$ from Equation 6.18 and then using Figure 6.17. Figure 6.17 has been calculated assuming:

- a ring of wells close enough to act as a single equivalent pumped well of radius re
- a uniform soil stratum with constant soil parameters $k_h$, $E'_o$ and $c_{hv}$
- purely horizontal flow
- no sources of vertical or horizontal recharge within the current radius of influence of the single equivalent well
- no well losses (seepage face effects).

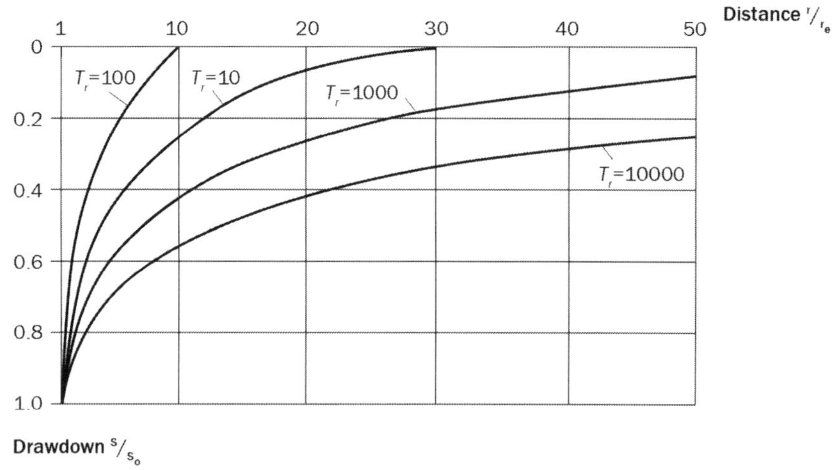


*Figure 6.17 Dimensionless drawdown curves for horizontal radial flow to a ring of wells acting as a single equivalent pumped well in a low permeability soil (after Powrie and Preene, 1994a)*

Finally, the pumping rates implied by the hydraulic gradients at entry into the equivalent slot or well shown in Figures 6.16 and 6.17 may be greater than those calculated using the methods described in Section 6.2 for steady-state conditions.

## 6.4.3 Rate of drawdown in moderate to high permeability soil

**See also**
Figure 6.8b Unconfined aquifer

In an unconfined coarse-grained soil of moderate to high permeability ($k$ greater than about $5 \times 10^{-5}$ m/s), the time taken to achieve the required drawdown may in principle be estimated from the pumping rate and the volume of water that must be released from storage. For plane flow to an excavation idealised as an infinite slot in an unconfined aquifer of initial saturated depth $H$, permeability $k$ and storage coefficient $S$, with a drawdown immediately outside the slot of $s_o = H/2$ and a distance of influence, $L_o$ (Figure 6.8b), the time $t$ taken to achieve steady-state conditions is given by:

Plane flow: $$t = \frac{8L_o^2 S}{27 s_o k} \qquad (6.19)$$

assuming pumping at the steady-state flow rate. Taking $L_o$ = 100 m, $s_o$ = 10 m, $S$ = 0.2 and $k = 10^{-4}$ m/s, Equation 6.19 gives $t \approx 7$ days. In reality, provided that the installed pumping capacity is greater than that needed at the steady-state (as is often the case), the time taken to achieve full drawdown in an unconfined coarse-grained aquifer does not seem to be a significant programming constraint, probably because:

- according to Equation 6.19, $t$ decreases rapidly with decreasing distance of influence, $L_o$, and increasing permeability, $k$
- owing to capillary effects, the storage coefficient, $S$, decreases significantly with decreasing grain size and hence permeability, $k$ (see Section 6.1.2).

In a confined coarse-grained aquifer in which the piezometric surface is not drawn down below the top of the aquifer, the soil will remain saturated. Any loss of water from the body of the aquifer is because of the compression of the soil skeleton resulting from the increase in vertical effective stress which accompanies a reduction in pore water pressure at constant total stress, according to Equation 1.2. Compression of the soil skeleton takes place at the same time as the pore water flows out of the soil, in the time-dependent process of consolidation (Section 6.6). The compression for a given increase in effective stress increases as the soil stiffness decreases, and the rate at which it occurs decreases with the soil permeability (which governs the ease with which water can flow out of the soil pores). In principle all soils consolidate, but the term is usually associated with soft, low permeability soils (ie clays and silts) because volume changes in stiff, high permeability soils (ie sands and gravels) are generally very small (because of their high stiffness) and occur very rapidly (because of their high permeability).

**See also**
6.6.2 Consolidation analysis

In addition to the soil stiffness in one-dimensional compression, $E'_o$, and the permeability, $k$, the time for consolidation depends on the maximum drainage path length, $d$ (see Equation 6.27, Section 6.6.2). Table 6.3 gives indicative order of magnitude times to achieve drawdown by consolidation for different soil types of high, moderate and low permeability, for a maximum drainage path length $d$ = 50 m. This shows that the time taken to achieve drawdown is often immaterial in fine sands and coarser soils, provided that the soil remains saturated (as will be the case in confined aquifers).

*Table 6.3 Indicative times for pore water pressure change by consolidation, with drainage path length of 50 m*

| Soil parameters* | Medium sand | Fine sand | Silt |
|---|---|---|---|
| Permeability $k$ (m/s) | $10^{-3}$ | $10^{-4}$ | $10^{-6}$ |
| Stiffness in one-dimensional compression $E'_o$ (MPa) | 100 | 50 | 10 |
| Time $t$ to achieve drawdown with drainage path length $d$ (= 50 m) | 4 minutes | 1.4 hours | 29 days |

**Note**

* Illustrative soil parameters at an average vertical effective stress of 100 kPa.

The time to achieve drawdown in a confined aquifer of moderate to high permeability can be estimated using the methods described in Section 6.4.2, provided that the aquifer remains confined at all locations during pumping. For horizontal plane flow to an equivalent slot, Figure 6.16 can be used in combination with Equation 6.20:

$$L_o = \sqrt{\frac{12kDt}{S}} \tag{6.20}$$

where $t$ is the elapsed time since pumping commenced, $D$ is the thickness of the confined aquifer, and all other terms are as defined previously. For horizontal radial flow to an equivalent well, Figure 6.17 can be used in combination with Equation 6.21:

$$T_r = \frac{kDt}{Sr_e^2} \tag{6.21}$$

An alternative approach to considering lines or groups of wells as equivalent slots or wells is to use the principle of superposition to calculate the drawdown at time $t$ from the cumulative effect of pumping from several wells simultaneously. This method is described in Sections 6.5.2 and 6.5.3.

# 6.5 ESTIMATION OF TIME-DEPENDENT DRAWDOWN PATTERN AROUND A GROUP OF WELLS

## 6.5.1 Groups of wells treated as an equivalent well or slot

The methods described in Section 6.4 assume the individual wells in a line or ring are closely spaced and can be modelled as equivalent slots or wells. In low permeability soils the drawdown pattern at time $t$ can be obtained for plane flow from Figure 6.16 and Equation 6.17 and for radial flow from Figure 6.17 and Equation 6.18. In soils of moderate to high permeability in confined aquifers, or for small drawdowns in unconfined aquifers, the drawdown pattern can be obtained from Figure 6.16 and Equation 6.20 for plane flow, and from Figure 6.17 and Equation 6.21 for radial flow.

## 6.5.2 Superposition analyses in confined aquifers

**See also**
5.3.1 Well pumping tests
6.3.1 Well losses
6.3.4 Well yields

If individual wells are widely spaced, it may not be appropriate to estimate the drawdown pattern by an equivalent well approach and a superposition method may be more suitable. This analysis uses the mathematical property of superposition applied to groundwater flow solutions for confined aquifers. In essence, superposition means that the drawdown at a given point from several pumped wells (at various distances apart) is equal to the sum of the drawdowns from each well taken individually (Figure 6.18). Complications arise in unconfined aquifers; because the saturated thickness reduces toward the wells, non-linearities are introduced and linear superposition is no longer valid (Section 6.5.3). Application of superposition is discussed further by Powers *et al* (2007). A detailed discussion including application to unconfined aquifers can be found in Bear (1979).

The drawdown is normally calculated at locations away from the pumped wells (eg beneath the deepest part of the proposed excavation). Calculating the drawdown inside each well in a groundwater control system is more difficult because well losses (Section 6.3.1) can be difficult to predict. If large well losses occur, the results of superposition analyses are less reliable, because the drawdown contribution from each well becomes uncertain.

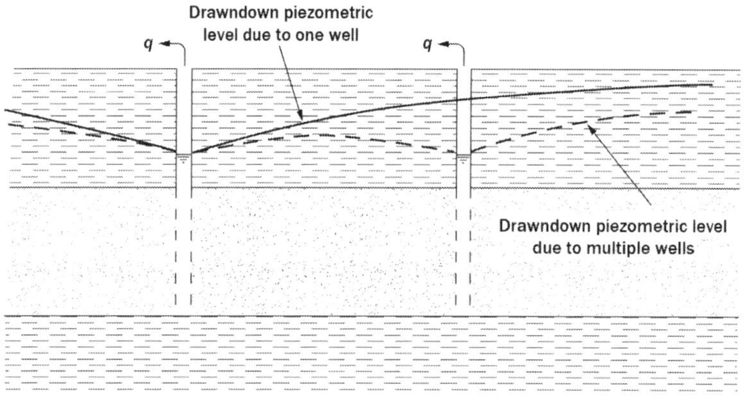


*Figure 6.18 Superposition of drawdown in a confined aquifer*

Superposition analysis, sometimes known as the cumulative drawdown method, can be used to predict the drawdown pattern around a group of wells or to calculate the flow rate required to achieve the target drawdown within an excavation. The cumulative drawdown $(H - h)$ at a given point in a confined aquifer from $n$ pumping wells can be expressed as the sum of the drawdown contributions $(H - h)$ from the individual wells each pumped at a flow rate $q$:

$$(H-h)=\sum_{1}^{n}(H-h)_i \tag{6.22}$$

For wells that fully penetrate a confined aquifer of isotropic permeability $k$, storage coefficient $S$ and thickness $D$, the drawdown contribution from each well (pumped at a constant flow rate $q$) at elapsed time t can be calculated using the method of Theis (1935). The resulting cumulative drawdown at a point is shown in Equation 6.23:

$$(H-h)=\sum_{1}^{n}\frac{q_i}{4\pi kD}W(u_i) \tag{6.23}$$

where $W(u)$ is the Theis well function (values of which are tabulated in most hydrogeological texts, eg Kruseman and De Ridder, 1990), $u = (r^2S)/(4kDt)$ and $r$ is the distance from each well to the point under consideration, and for small values of $u$, Equation 6.23 can be expressed as the Jacob formulae (Cooper and Jacob, 1946):

$$(H-h)=\sum_{1}^{n}\frac{q_i}{4\pi kD}\left(-0.5772-\ln\left[\frac{r_i^2S}{4kDt}\right]\right) \tag{6.24}$$

Kruseman and De Ridder (1990) indicate that Equation 6.24 is valid for $u < 0.1$, a condition which in many aquifers is satisfied after a few hours pumping and so can generally be used for the analysis of groundwater control systems.

For conditions not satisfying the assumptions of Equations 6.23 and 6.24 (ie isotropic confined aquifer, fully penetrating well pumped at constant flow rate) the drawdown contributions should be calculated using alternative formulae. Kruseman and De Ridder (1990) give solutions for a number of cases, including partially penetrating wells, anisotropic permeability, variable pumping rates and leaky aquifers.

The superposition method can be used to determine the drawdown pattern around a proposed group of wells or, by iteration, to estimate the number, yield and layout of wells to achieve the target drawdown. The method can be applied numerically (King, 1984), for example using routines written for spreadsheet programs to calculate the cumulative drawdown using Equations 6.23 or 6.24. Appropriate routines can calculate drawdown at various locations across the site and graphics packages can be used to produce contours of groundwater levels or drawdown. The method can also be used to estimate the time–drawdown relationship (Section 6.4) by calculating the cumulative drawdown at various times after pumping commences.

Results of superposition analyses depend on the chosen parameter values (principally permeability and storage coefficient). Ideally these should be determined from an appropriately analysed pumping test (Kruseman and De Ridder, 1990). If a pumping test has not been carried out and parameter values have not been determined sufficiently accurately by other means (eg inverse numerical modelling), the results of superposition analyses should be treated with caution.

If results of well pumping tests (Section 5.3.1) are available, the variation of drawdown with distance from the well recorded at time $t$ during the test can be used in a graphical cumulative drawdown method. This is based on the Jacob method (Kruseman and De Ridder, 1990), which uses Equation 6.24 expressed as:

$$(H-h)=\sum_{1}^{n}\frac{q_i}{2\pi kD}\ln\left[{R_o}/{r_i}\right] \tag{6.25}$$

where all terms are as defined previously, apart from $R_o$, which is the distance of influence at time $t$. In practice Equation 6.25 is often evaluated not numerically, but graphically from the pumping test results, without the need for complex mathematics. A superposition method to determine the number, yield and layout of wells to achieve the target drawdown in the required areas is described as follows:

1 Based on the depth of excavation and initial groundwater level, determine the target drawdown in certain key areas of the excavation. These might include the centre and corners of the excavation.

2 From the drawdown data at the end of the pumping test, construct a drawdown vs. log distance plot on semi-logarithmic axes (Figure 6.19a).

3 Convert the drawdown data to specific drawdown (drawdown per unit flow rate) by dividing the drawdown by the steady-state flow rate recorded in the test. A straight line should be drawn through the piezometer data to produce the specific drawdown plot to be used in design (Figure 6.19b).

4 Draw a plan showing the proposed well locations and the positions where drawdown is to be calculated, and measure the distances from each well to the drawdown calculation locations.

5 Estimate or determine the proposed well yields (either using the methods of Section 6.3.4 or based on the pumping test results).

6 For each drawdown location, use the specific drawdown plot to determine the contribution from each well. The actual drawdown contributed by each well is calculated by multiplying the specific drawdown for each well by the proposed flow rate for that well. The total drawdown at each calculation point is the sum of the drawdown contributions from each well (Box 6.10). In practice, observed drawdowns are sometimes rather less than those calculated directly by this method. Box 6.10 shows data where the observed drawdown was 92 per cent of the calculated value. The reduced drawdown may be a result of interference between closely spaced wells (see below). In some circumstances the calculated cumulative drawdown is multiplied by an empirical superposition factor (examples of the range of possible values are given below).

7 If the drawdown is insufficient, rearrange the wells or add to the capacity of the system (by adding wells or increasing individual well capacity) and repeat the analysis.

## Interference between wells

Cumulative drawdown analysis assumes that the wells do not interfere significantly with each other in terms of yield and influence or drawdown. For wells installed at relatively wide spacing (> 20 m) in confined aquifers, and where the aquifer remains confined after drawdown, interference is usually low. The observed drawdowns may be close to those predicted directly from superposition analyses. This is demonstrated by the case study in Box 6.10, where observed drawdowns were 92 per cent of superposition calculations. In confined aquifers, superposition of cumulative drawdowns of 80 per cent or more may be assumed in design. To allow for this, the results of superposition calculations can be multiplied by an empirical superposition factor of between 0.95 and 0.8.

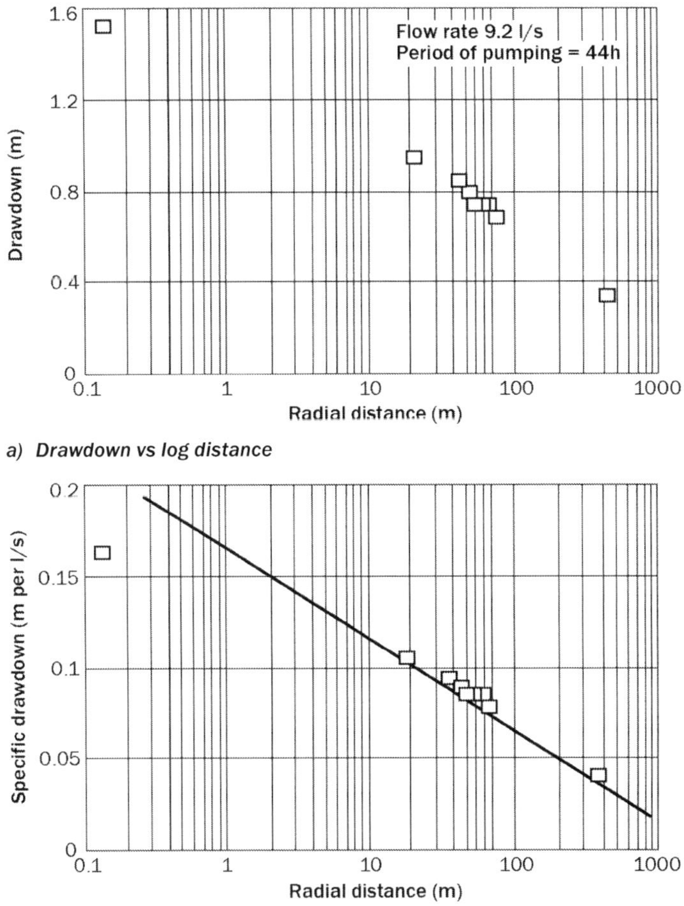


*a) Drawdown vs log distance*

*b) Specific drawdown vs log distance*

*Figure 6.19 Drawdown–log distance relationships for pumping tests*

*Box 6.10 Case study of superposition calculation using pumping test data*

Pumping from a system of deepwells in a confined chalk aquifer.

Estimate drawdown in observation well 8 (specific drawdown determined from single well pumping test; data given in Figure 6.19b).

| Well | Flow rate (l/s) | Distance to well 8 (m) | Specific drawdown (m per l/s) | Calculated drawdown (m) |
|---|---|---|---|---|
| 1 | 8.5 | 82 | 0.079 | 0.67 |
| 2 | 8.5 | 100 | 0.072 | 0.60 |
| 6 | 11.0 | 50 | 0.082 | 0.91 |
| 7 | 11.0 | 20 | 0.103 | 1.13 |
| | | Total calculated drawdown at well 8 = 3.31 m | | |

Actual drawdown recorded at well 8 after 44 hours was 3.06 m.

Therefore drawdown achieved is 3.06/3.31 = 92 per cent of calculated cumulative drawdown (after Preene and Roberts, 1994).

## 6.5.3 Superposition analyses in unconfined aquifers

If the aquifer is unconfined, or a confined aquifer becomes locally unconfined, some interference is unavoidable and a reduced percentage superposition should be applied. The saturated thickness decreases as drawdown increases, making each additional well less efficient than the initial wells. Despite the principle of superposition not being valid for unconfined aquifers, the method has been used for unconfined aquifers where the reduction in aquifer thickness by drawdown did not exceed about 20 per cent. Outside those drawdown limits, the method has been applied and empirical superposition factors of 0.6 to 0.8 have been used.

# 6.6 ESTIMATION OF SETTLEMENTS

## 6.6.1 Mechanisms of settlement

The main aim of a groundwater control system is to reduce pore water pressures in the soil surrounding an excavation, so that the sides and base of the excavation remain stable. The vertical total stresses in the soil outside the excavation will usually remain unchanged, so that the reduction in pore water pressure must (according to Equation 1.2) be accompanied by an increase in vertical effective stress. This will cause a vertical strain or settlement of the soil.

Many of the soils that are suitable for dewatering (such as sandy gravels) are comparatively permeable and stiff, so the ground movements, which result from the changes in pore water pressure and effective stress, usually occur very quickly and are unnoticeably small. However, where softer soils are present (for example, as an overlying layer of alluvial clay, silt or peat), there may be concern that settlement of the soil could damage nearby buildings and buried services. As softer soils, with the exception of some peats, are generally less permeable, the settlements that occur as the soil consolidates may take some time to develop.

A second possible cause of ground movements associated with dewatering systems is the movement of soil particles. This can occur if the well screens and filters are inappropriate for the ground conditions, allowing the continued removal of fine particles. Surface settlements from the continued removal of fine soil particles with the pumped groundwater are generally localised, but potentially large and serious: they must therefore be prevented. Ground movements as a consequence of loss of fines can also occur in passive drainage systems (eg French drains and pipe bedding layers) that have not been installed in compliance with the filter rules (Section 6.3). Rowe (1986) gives two examples of problems of this type.

**See also**
Case study 7.6

A third possible cause is that the pore water pressure reduction achieved by the dewatering system may be insufficient to prevent instability, perhaps because of features such as high permeability lenses or shoestrings which were not identified at the site investigation stage (see Case study 7.6).

Soil settlement as a result of loss of fines or insufficient reduction of pore water pressure should not occur with a groundwater control system that has been properly designed and installed and for which an adequate site investigation (Chapter 5) has been carried out.

## 6.6.2 Settlement from increase in vertical effective stress

### Theoretical basis

The vertical settlement $\rho$ of a uniform layer of soil of thickness $D$ and stiffness in one-dimensional compression $E'_o$ subjected to a uniform increase in vertical effective stress $\Delta\sigma'_v$ may be calculated using Equation 6.26:

$$\rho = \frac{D\Delta\sigma'_v}{E'_o} = \frac{D\gamma_w s}{E'_o} \qquad (6.26)$$

Assuming that the vertical total stress remains constant, the increase in vertical effective stress $\Delta\sigma'_v$ is equal to the reduction in pore water pressure $\Delta u$, which is in turn equal to the unit weight of water $\gamma_w$ multiplied by the drawdown $s$. Equation 6.26 shows that the magnitude of the settlement $\rho$ increases with the thickness of the soil layer $D$ and the drawdown $s$, and decreases as the one-dimensional stiffness $E'_0$ increases. Increases in effective stress from reductions in pore water pressure occur only within the distance (or radius) of influence $L_0$ (or $R_0$) of the dewatering system, so that the magnitude of $L_0$ may also be relevant.

## Soil parameters and factors necessary to assess settlement

The key parameters in assessing the potential for settlements resulting from the operation of a groundwater control system are:

- the drawdown, $s$, or reduction in pore water pressure
- the thickness(es), $D$, of the soil layer(s) affected
- the soil stiffness(es) in one-dimensional compression, $E'_0$, or the coefficient(s) of volume compressibility, $m_v$
- the distance (or radius) of influence of the dewatering system, $L_0$ (or $R_0$).

In practice there may be more than one soil layer present. Also, the soil stiffness and the increase in effective stress, which results from the reduction in pore water pressure caused by the dewatering system, will probably vary with depth. In any of these cases the soil should be considered as a number of layers, each of which is characterised by a uniform stiffness in one-dimensional compression $E'_0$ and a uniform increase in vertical effective stress $\Delta\sigma'_v$. The surface settlement at any point is the sum of the compression of each individual layer. Even if there is only one soil type present, this procedure can be used to take account in a stepwise fashion of an increase in soil stiffness (or a variation in vertical effective stress increment) with depth.

The time $t$ taken for the reduction in pore water pressure (and hence the settlement) from the operation of the groundwater control system to take effect depends on the consolidation characteristics of the soil. In one-dimensional vertical compression, the time $t$ for settlement to be completed is given approximately by Equation 6.27, with the dimensionless time factor $T = 1$:

$$T = \frac{c_v t}{d^2}, \text{ where } c_v = \frac{kE'_o}{\gamma_w} \qquad (6.27)$$

where $c_v$ is the consolidation coefficient and $d$ is the maximum drainage path length. Thus if the time over which settlement may occur is important, the soil permeability, $k$, and the maximum drainage path length, $d$, also have to be assessed.

The soil stiffness in one-dimensional compression $E'_0$ is not a constant, but depends on many factors, including the stress history or density of the soil, the current stress and the changes in stress to which the soil will be subjected. It is important that numerical values of $E'_0$ are determined in an appropriate way. For example, it is easy to underestimate the soil stiffness because the changes in stress and strain associated with groundwater control are likely to be small, and the stiffness of a soil can be large at small strains. Some common methods of estimating soil stiffness are summarised in Table 6.4. Only the oedometer test gives the one-dimensional stiffness $E'_0$ directly; the other tests give the shear modulus, $G$, or the Young's modulus, $E$, which are related to $E'_0$ by Poisson's ratio, ν' (see Powrie, 2013). The range of values of $E'_0$ that occur in various soil types can be estimated from Table 6.5, which gives approximate ratios between soil stiffness in one-dimensional compression and vertical effective stress. The soil stiffness values presented in Table 6.5 are indicative values only, in the absence of site-specific data. Their suitability for use on a given site should be assessed during the analysis of dewatering-related settlement.

*Table 6.4 Common methods of estimating soil stiffness*

| Method | Comments | Reference |
|---|---|---|
| Oedometer test | Laboratory test<br>Sample size, soil fabric and sample disturbance may affect results (Rowe, 1972) | Powrie (2013) |
| Triaxial test | Laboratory test<br>Sample size, soil fabric and sample disturbance may affect results (Rowe, 1972) | Powrie (2013) |
| Plate bearing test | *In situ* method<br>Thickness or volume of soil tested may be too small | Clayton *et al* (1995) |
| Standard penetration test | *In situ* method<br>Empirical correlation | Clayton (1995) |
| Cone penetrometer test | *In situ* method<br>Empirical correlation | Robertson and Campanella (1983), Meigh (1987) |
| Pressuremeter test | *In situ* method<br>Soil is loaded in the horizontal, rather than the vertical, direction | Mair and Wood (1987) |

*Table 6.5 Approximate ratios between soil stiffness in one-dimensional compression and vertical effective stress for typical soils*

| Indicative soil type | Ratio of stiffness in one-dimensional compression $E'_o$ to vertical effective stress $\sigma'v$ $E'_o/\sigma'v$ |
|---|---|
| Dense sand, recompression (overconsolidated)<br>Dense sand, first compression (normally consolidated) | 600<br>2000 |
| Loose to medium density sand, recompression (over-consolidated)<br>Loose to medium density sand, first compression (normally consolidated) | 500<br>150 |
| Stiff overconsolidated clay | 400 |
| Soft normally consolidated clay | 20 |
| Peat | 10 |

**See also**
5.3 Permeability testing
6.1.2 Permeability selection

Values of the consolidation coefficient, $c_v$, should be determined with care. For example, a value measured in an oedometer test with vertical drainage is likely to underestimate the speed of consolidation in a layered soil in the field if the dominant direction of drainage is horizontal. An indirect approach is sometimes used to estimate $c_v$ (see Al-Dhahir *et al*, 1969), using soil stiffness values obtained from oedometer or triaxial tests and coefficients of permeability from *in situ* tests (see Sections 5.3 and 6.1.2).

Using unsuitable values of soil stiffness to estimate dewatering-induced settlements can cause unnecessary concern. In particular, simple empirical correlations between soil stiffness and standard penetration test (SPT) blow count or static cone penetrometer resistance are generally based on the back-analysis of shallow foundations, for which lower soil stiffness values are appropriate because of the larger strains involved. If stiffnesses from these correlations are used, settlements from groundwater control may be overestimated.

Box 6.11 shows basic settlements, calculated according to Equation 6.26, for different values of stiffness in one-dimensional compression $E'_o$.

*Box 6.11 Basic settlements for soils of different stiffness in one-dimensional compression*

The basic settlement is defined as the compression of a soil layer 1 m thick from an increase in vertical effective stress corresponding to a drawdown of 1 m. For a given situation, the total settlement in mm may be obtained by multiplying the basic settlement by the drawdown and the thickness of the soil layer (both in metres)

| One-dimensional soil stiffness, $E'_0$ (MPa) | 1 | 5 | 10 | 15 | 20 | 25 |
|---|---|---|---|---|---|---|
| Basic settlement (mm) | 10.0 | 2.0 | 1.0 | 0.667 | 0.5 | 0.4 |
| One-dimensional soil stiffness, $E'_0$ (MPa) | 40 | 50 | 75 | 100 | 150 | 200 |
| Basic settlement (mm) | 0.25 | 0.20 | 0.133 | 0.10 | 0.067 | 0.05 |

## Coarse-grained soils and over consolidated clays

Experience shows that most medium dense or denser coarse-grained soils (ie sands, gravels) and heavily over consolidated clays (eg Glacial Till or London Clay) are sufficiently stiff to accommodate the increases in effective stress likely to result from dewatering without significant settlement.

For an overconsolidated sand, where $E'_0$ might be approximately 200 MPa, Box 6.11 suggests a settlement of only 0.05 mm per metre drawdown per metre thickness, giving a settlement of 2.5 mm for an average drawdown of 5 m over a soil layer 10 m thick. For a more compressible sand with $E'_0$ = 20 MPa, the corresponding settlement is 0.5 mm per metre drawdown per metre thickness, or 25 mm for an average drawdown of 5 m over a soil layer 10 m thick.

## Fine-grained and normally consolidated soils

In practice, significant settlements are most likely to occur when a soft, normally consolidated stratum (such as alluvial clay, silt or peat) is subjected to an increase in vertical effective stress. This may result from the underdrainage of a permeable layer (see below and Box 6.13) or from pumping directly from the fine-grained stratum using vacuum-assisted wells.

Large settlements can be expected in such soft soils. For a soft silty clay, where $E'_0$ might be of the order of 2 MPa, Box 6.11 suggests a settlement of 5 mm per metre drawdown per metre thickness, giving a settlement of 250 mm for an average drawdown of 5 m over a soil layer 10 m thick.

## Settlements due to other construction activities

Settlements resulting from groundwater control may or may not be significant compared to the settlements that might be expected to result from other construction activities, for example:

- sheet-pile or diaphragm wall installation: settlements may be up to 0.2 per cent of the depth of the wall, ie 40 mm for a wall 20 m deep (Clough and O'Rourke, 1990)
- excavation in front of a sheet-pile or diaphragm wall: settlements may be up to one per cent of the excavated depth in sand and soft to hard clay, ie 100 mm for an excavation 10 m deep (Peck, 1969b).

Nevertheless, settlements resulting from groundwater control are additional to the settlements caused by other construction activities, and may be of sufficient lateral extent to affect existing structures not influenced by other construction activities. The effect of other construction activities is illustrated by the case study described in Box 6.12, in which significant settlement occurred before groundwater control was commenced.

*Box 6.12 Case study of settlements caused by excavation and groundwater control*

A large excavation was constructed adjacent to an existing embankment. The sides of the excavation were supported by sheet-piles propped against 'dumplings' (mounds of earth) left in place within the excavation. Ground conditions consisted of 3 m of firm silty clay overlying medium dense sands, with groundwater levels close to original ground level. An ejector well system was used to lower the groundwater levels by approximately 10 m, and ground anchors were installed as part of the permanent works. Site measurements (shown below) indicate that settlements of the order of 40 mm occurred before pumping began – significantly more than the 10 mm to 15 mm of settlement recorded during the first month of pumping. The pre-pumping settlements may have been caused by the installation of the sheet-piling and some initial shallow excavations made above groundwater level.

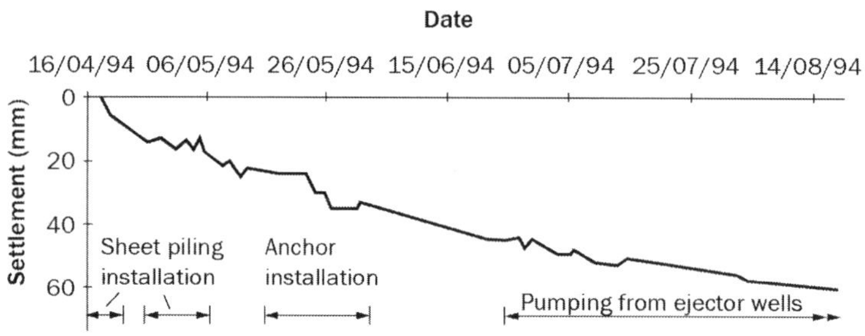


***Settlements from groundwater control and other construction activities***

On completion of groundwater control, pore water pressures will recover to their original levels (or to equilibrium with any permanent drainage that has been installed). As a result, effective stresses may decrease, halting settlement once groundwater levels fully recover and possibly inducing some swelling or heave of the soil back toward original ground levels.

## Underdrainage of a compressible stratum

Settlements caused by dewatering are likely to be a problem when pumping from a confined aquifer overlain by a compressible stratum such as soft clay or peat, even though the aquifer itself has a high stiffness. The compressible layer, although not pumped directly, will consolidate because the drainage of pore water downward into the underlying aquifer causes an increase in vertical effective stress. A case study involving settlements caused by pumping water from an aquifer overlain by a compressible stratum of lower permeability is given in Box 6.13.

*Box 6.13 Case study of dewatering-induced settlements caused by the underdrainage of a compressible layer*

**The problem**

Wellpoints were used to lower the water table from an initial level of 0.3 m bgl to 4.3 m bgl for a series of small excavations within an area less than 30 m square in plan. Ground conditions comprised approximately 4 m of topsoil, peat and soft alluvial clay underlain by a glacial sand and gravel aquifer (see figure). After about three weeks pumping, owners of properties up to 500 m away began to complain of structural damage, and the dewatering system was switched off.

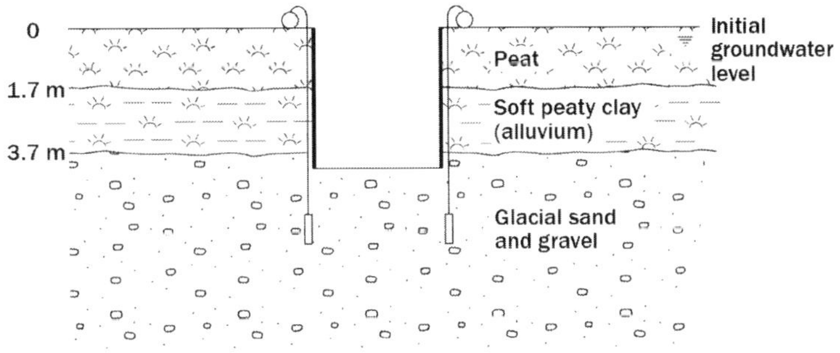


*Ground conditions*

**The explanation**

The groundwater level in the sand and gravel aquifer was lowered quite quickly, following which the compressible alluvial clay and peat began to consolidate by vertical drainage of pore water down into the sand and gravel. A long-term soil surface settlement of about 150 mm was subsequently calculated from Equation 6.26; values for the one-dimensional stiffness $E_o$ (measured over appropriate stress increments in oedometer tests) were 0.5 MPa for the clay and 0.2 MPa for the peat layers. An analysis in which the clay and the peat were treated as a single layer suggested a surface settlement of over 80 mm after twenty days, assuming an effective vertical permeability of $10^{-8}$ m/s. The distance from the excavation to some of the properties alleged to have suffered settlement damage is explained by the piezometric levels in the sand and gravel aquifer at various times after pumping had ceased, which showed very little variation with horizontal distance up to 250 m away from the excavation (see figure). The furthest property allegedly affected by settlement coincided exactly with the edge of the peat deposit indicated on the geological map of the area.

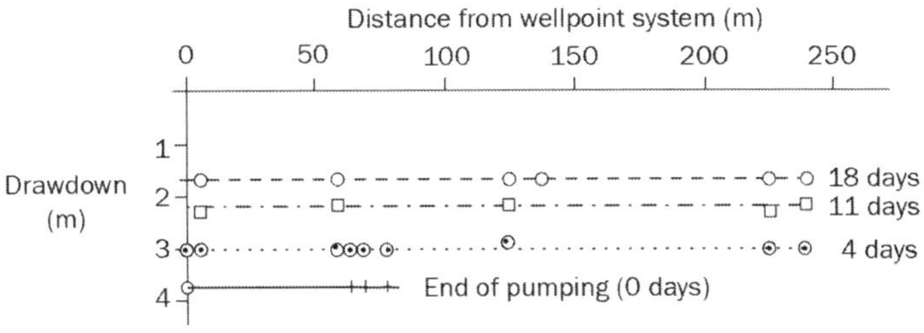


*Piezometric levels in the sand and gravel aquifer at various times after pumping stopped (after Powrie, 2013)*

## Differential settlements

**See also**
Case study 7.8

In general, damage to buildings is more likely to arise from differential rather than uniform settlement. Guidelines developed by Burland and Wroth (1975) and others can be used to estimate maximum acceptable values for differential settlement for a building of given construction, in order to avoid certain types of damage (see Powers, 1985).

In the case study described in Box 6.13, settlements occurred because of the consolidation of a low permeability layer by vertical drainage into an underlying aquifer from which groundwater was being pumped. If the compressible layer had been homogeneous and of uniform thickness, these settlements should in theory have developed at the same rate over a wide area. In reality, uniform conditions are not common and differential settlements are likely to occur if:

- the compressible strata vary in thickness
- the foundations of the building have not been designed to a consistent load factor (for example, a building that has been partly underpinned, or where there are piles under part of the building only, see Case study 7.8)
- the drawdown varies significantly with distance beneath the building (ie the cone of depression is steep or the building is of very large plan area).

The effects of all of these are likely to be magnified if the stiffness of the soil is low. Powers *et al* (2007) cites the presence of a compressible stratum as the most significant cause of settlement damage to buildings resulting from groundwater control operations; factors such as the magnitude of the drawdown and a variation in foundation type and loading are often of only secondary importance.

In the case study described in Box 6.13, the ground conditions across the site were very variable. Six of the boreholes indicated thicknesses of between 0.6 m and 2 m for the peat, and between 0 m and 2 m for the soft clay layer. In two further boreholes towards the edge of the site, neither stratum was present. Also, one of the properties allegedly affected was a supermarket, whose car park occupied the site of a former industrial building. Uniform settlements might not have been a problem if the piled foundations of the old building had not been left in place beneath the surface of the car park. In the event, the settlement of the surrounding ground resulted in an unsightly array of humps in the surface of the car park at the location of each pile.

## Pumping directly from a compressible stratum

**See also**
Figure 6.20 Drawdown vs. distance for plane flow

In cases where a pumped well system is installed to control the pore water pressures in a fine-grained soil and there is no underlying more permeable layer, consolidation will occur as the pore water is drawn towards the pumped well system in horizontal flow (Figures 6.16 and 6.17). In these circumstances, the drawdown at any time (and hence the increase in vertical effective stress) varies with distance from the pore water pressure control system. Differential settlements must therefore be expected, even in a homogeneous stratum of uniform thickness. The rate of settlement is controlled by the stiffness in one-dimensional vertical compression, $E'_o$, and the horizontal permeability, $k_h$, of the soil. Settlements cannot be prevented because the purpose of the pumped well system is to reduce pore water pressures in the compressible stratum. As the settlement depends on the drawdown, differential settlements are related to the slope of the distance–drawdown curve (eg Figures 6.16 and 6.17). Provided that the slope of the drawdown curve is shallow, the soil is reasonably stiff and the structure at risk is small in scale compared with the area affected by drawdown, differential settlements are likely to be small.

In summary, the soil settlements induced by dewatering will in many soils be small, particularly in comparison with those caused by other construction activities such as excavation in front of a sheet-pile retaining wall. However, if there are thick layers of compressible soils (such as alluvial clays, silts and peats), dewatering settlements may be more significant. In such cases, soil movements can be estimated using the relatively simple effective stress methods described in this section. The fact that consolidation is time-dependent should also be taken into account. The parameters used to calculate settlements must be appropriate to the stress and state of the soil, and the changes in stress to which it is likely to be subjected.

# 6.7 KEY REFERENCES

ANDERSON, M P and WOESSNER, W W (2015) *Applied groundwater modelling, second edition*, Academic Press Inc, New York, USA (ISBN: 978-0-12058-103-0)

BOND, A (ed) (1994) *Validation and use of geotechnical software*, Association of Geotechnical and Geoenvironmental Specialists (AGS), Beckenham, Kent (ISBN: 978-0-95192-714-4)

CASHMAN, P M and PREENE, M (2012) *Groundwater lowering in construction: a practical guide to dewatering, second edition*, CRC Press, Boca Raton, USA (ISBN: 978-0-41921-110-5)

DRISCOLL, R, SCOTT, P and POWELL, J (2008) *EC7 – implications for UK practice*, C641, CIRIA, London (ISBN: 978-0-86017-641-1). Go to: **www.ciria.org**

HAUSMANN, M R (1990) *Engineering principles of ground modification*, McGraw-Hill, New York, USA (ISBN: 978-0-07027-279-8)

KRUSEMAN, G P and DE RIDDER, N A (1990) *Analysis and evaluation of pumping test data, second edition*, Publication 47, International Institute for Land Reclamation and Improvement, Wageningen, The Netherlands (ISBN: 978-9-07075-420-4)

MANSUR, C I and KAUFMAN, R I (1962) "Dewatering". In: G A Leonards (ed) *Foundation engineering*, McGraw-Hill, New York, pp 241–350

MISSTEAR, B, BANKS, D and CLARK, L J (2006) *Water wells and boreholes*, Wiley, Chichester (ISBN 978-0-470-84989-7)

POWERS, J P (1985) *Dewatering – avoiding its unwanted side effects*, American Society of Civil Engineers, New York, USA (ISBN: 978-0-87262-459-7)

POWRIE, W and PREENE, M (1992) "Equivalent well analysis of construction dewatering systems" *Géotechnique*, vol 42, 4, Institution of Civil Engineers, London, pp 635–639

POWRIE, W and PREENE, M (1994a) "Time–drawdown behaviour of construction dewatering systems in fine soils" *Géotechnique*, vol 44, 1, Institution of Civil Engineers, London, pp 83–100

SHERARD, J L, DUNNIGAN, L P and TALBOT, J R (1984a) "Basic properties of sand and gravel filters" *Journal of Geotechnical Engineering*, vol 110, 6, American Society of Civil Engineers, Reston, USA, pp 684–700

## Standards

BS EN 1997-1:2004 *Eurocode 7: Geotechnical design. Part 1: General rules*

# 7 From design to practice

## 7.1 INTRODUCTION

**See also**
Chapter 6 Design
Figure 6.1 Design

Chapter 6 presents methods of analysis to allow estimation of total flow rate, well yields, time to achieve drawdown and potential settlements. The principal stages in design are shown in Figure 6.1. However, to move from these results to a groundwater control system on site involves judgements based on the experience of the designer, and on practical and economic considerations. This section presents some case studies illustrating the transition from theory to practice. Several of these describe projects where all did not go according to plan. In fact such cases are quite rare (where there has been adequate planning and investigation), but problems encountered in practice can illuminate specific lessons.

Experience has shown that where groundwater control systems perform poorly, the root cause is rarely simply incorrect calculations, or even errors in permeability selection; the problem often arises from an inappropriate conceptual model – getting the idea wrong. 'Inadequate site investigation' is commonly cited as the reason for an incorrect conceptual model, but it may also arise from poor interpretation of the groundwater risks when formulating the model. Designers may be tempted to fit the ground conditions to match their model, in which case the groundwater control is unlikely to be successful.

Different groundwater control methods have a wide range of application, as shown in Figure 7.1. If the required drawdown and approximate soil permeability are known, an initial assessment can be made of the appropriate groundwater control technique by finding the corresponding point on Figure 7.1. The shaded areas of this diagram show where the techniques overlap and one may be used in place of the other.

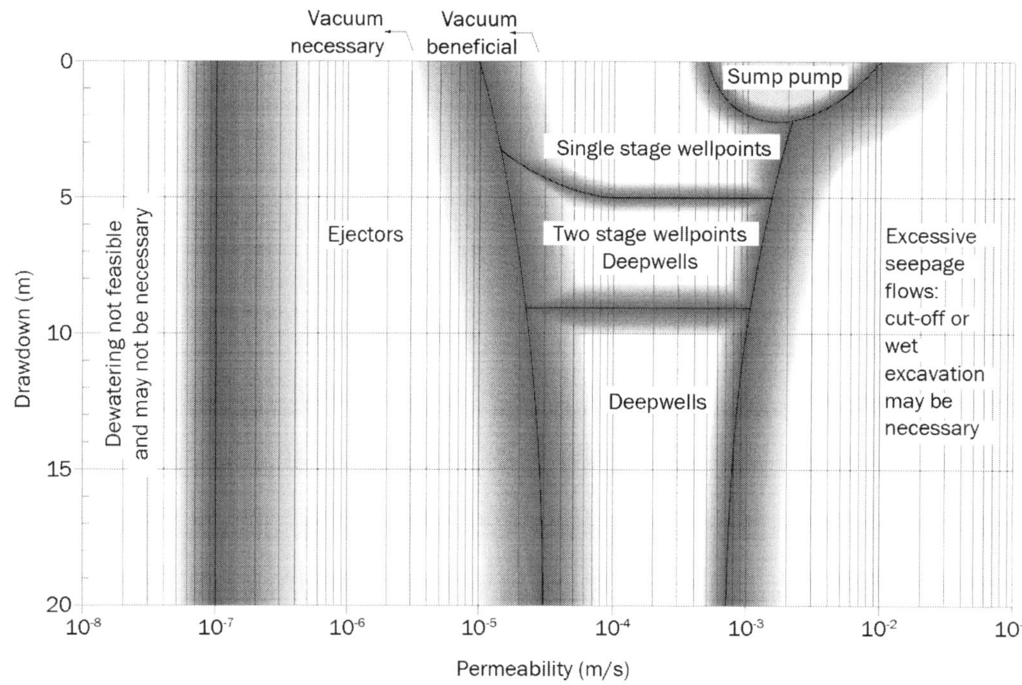


*Figure 7.1 Range of application of pumped well groundwater control techniques (adapted from Roberts and Preene, 1994a, and modified after Cashman, 1994b)*

# 7.2 THE OBSERVATIONAL METHOD

Even when thorough site investigations are carried out, in some circumstances the complexity of the ground conditions may mean that the design of a groundwater control system cannot be finalised, other than very tentatively. One solution sometimes adopted is to proceed by the Observational Method originally proposed for geotechnical engineering by Peck (1969a). Nicholson (1994) states that:

> *The method provides a way of controlling safety while minimising construction costs, so long as the design can be modified during construction. Peck identified two applications for the Observational Method:*
>
> *a ab initio: from inception of the project*
>
> *b best way out: during construction when unexpected site problems develop.*
>
> *Peck's observational method involves developing an initial design based on the most probable conditions, together with predictions of behaviour. Calculations based on the most unfavourable conditions are also made and are used to identify contingency plans and trigger values for the monitoring system. Peck proposed that the construction work should be started using the most probable design. If the monitoring records exceed the predicted behaviour, then the predefined contingency plans would be triggered. The response time for monitoring and implementation of the contingency plan must be appropriate to control the work.*

**See also**
3.4 Monitoring

Groundwater control systems are suitable for the Observational Method (as illustrated in Box 7.1) because they can easily be modified (eg by the addition of extra wells or by changing pump sizes) and are easy to monitor (see Section 3.4). Further examples are given in Roberts and Preene (1994b) and in Nicholson *et al* (1999).

The *ab initio* method tends to be applied to large projects or where the main contract is design and build and the groundwater control requirements may not be finalised until late into the project. The method can allow fine-tuning of the number of wells required and there may be a temptation to install only the bare minimum necessary to achieve the drawdown. This temptation should be avoided, because it is also important to consider the need for standby plant, alarm facilities and the potential for chemical or bacterial clogging (see Section 3.4) to be sure that drawdowns will be maintained during the construction period. The best way out method is often used to plan the uprating or modification of a system that is performing poorly; in effect the initial dewatering system is monitored and considered as a trial or large-scale pumping test.

*Box 7.1 Case study of the use of the observational method (after Roberts and Preene, 1994b)*

A pumping station required a 10 m drawdown in a glacial sand and gravel stratum – described on the borehole logs as silty sand and gravel with abundant cobbles and boulders. The PSD data indicated a permeability range of $10^{-6}$ m/s to $10^{-2}$ m/s, which covers most methods of dewatering and extends well into the zone requiring a physical cut-off on Figure 7.1. As silt and sand-size particles were largely absent from some of the samples, loss of fines during sampling was suspected. A pumping test had been carried out but, because only small flow rates and small drawdowns were achieved, results were inconclusive. An initial array of 20 ejector wells was installed but achieved only part of the necessary drawdown. Analysis of individual well flow rates and drawdowns in piezometers revealed that drawdowns were much less at one end of the site than at the other, despite the site being only 30 m by 20 m in plan. The system was uprated on the basis of this analysis; an additional 17 ejector wells and 7 deepwells were installed, and achieved the required drawdown. Most of the additional wells were installed at the end of the site where the unfavourable high flow rate–low drawdown conditions occurred.

30 m

| | | | |
|---|---|---|---|
| Number of deep wells | 0 | 2 | 5 |
| Number of ejectors | 11 | 8 | 18 |
| Total flow | 3.3 l/s | 5.2 l/s | 13.4 l/s |

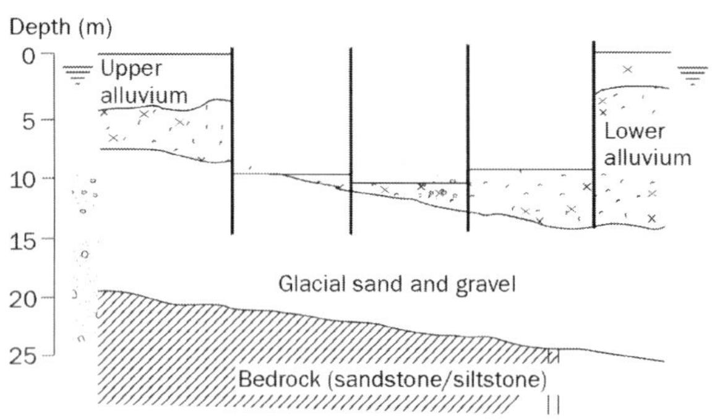


*Excavation cross-section*

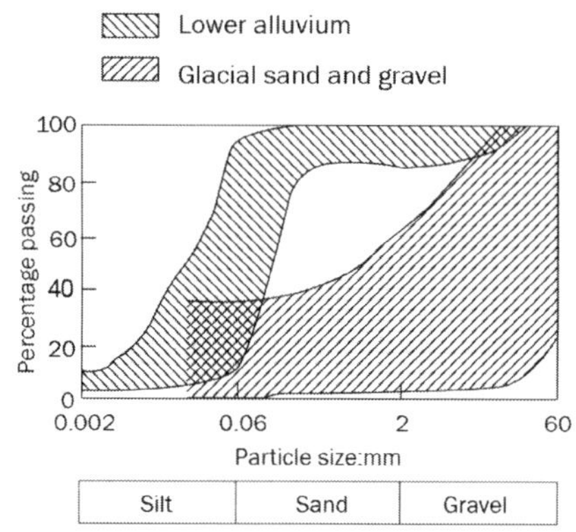


*Soil grading envelopes*

Back-analysis of the completed system suggested that the reduced initial drawdowns at one end of the site were probably the result of a boundary condition effect such as a close source of recharge or change in thickness of the aquifer, rather than a simple variation in permeability.

# 7.3 CASE STUDIES

## Case study 7.1 Use of deepwells instead of wellpoint system

### Background

See also
6.1 Groundwater modelling

An appropriate conceptual model (see Section 6.1) should allow the inter-relationship between groundwater flow in the various strata at a site to be identified. This then influences the choice of groundwater control method.

### Case history

A series of several shallow excavations to 5 m depth were to be dug over an area of approximately 150 m by 100 m as part of a new sewage treatment works. Ground conditions at shallow depth were fill and fine sand with groundwater levels at 1 m to 2 m bgl. Because the excavations were shallow, a wellpoint system was considered initially, but rings of wellpoints would have been needed around each excavation, both restricting access and increasing running costs. From the site investigation data a relatively permeable sandy gravel layer was identified at 10 m to 12 m depth. This was included in the conceptual model shown below and a dewatering scheme was designed with deepwells penetrating to the gravel layer. These wells were much deeper than the wellpoints would have been, but the aim was to lower the piezometric level in the gravel over a wide area and then let the overlying sands drain down into the gravel – a method known as underdrainage. In the event eight deepwells were used.

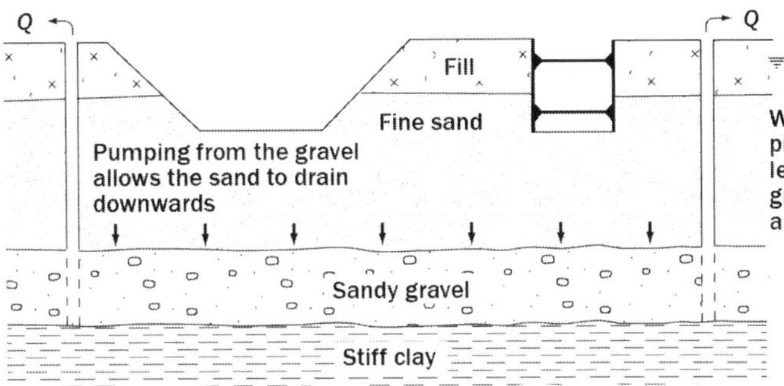


*Figure 7.2 Use of deep gravel layer to underdrain overlying finer soils*

### Comment

A degree of lateral thinking and the development of a conceptual model that recognised the presence of a deep permeable layer suitable for underdrainage enabled groundwater to be controlled using a small number of deepwells. This was more cost-effective than the obvious solution of large numbers of wellpoints. Installation costs of the two methods were similar but the deepwell option had the advantage of lower running costs over the project period. Also, the deepwell option imposed fewer access restrictions on the excavation contractor compared to the wellpoint solution (where headermains would have been laid around each excavation).

# Case study 7.2 Excessive flow rates in very permeable soil

## Background

**See also**

Box 6.1 Sensitivity analysis

At the higher end of the permeability range, very large flow rates can make dewatering unfeasible. The flow rate will be roughly proportional to permeability, so if the permeability used in design is in error by, say, 50 per cent (which is fairly likely), the actual flow rate will increase by about the same amount. In a fine sand where the flow rate might be 5 or 10 l/s, a doubling of the flow rate is unlikely to be a major problem. However, in a very permeable gravel ($k > 10^{-3}$ m/s), the design flow rate might be several hundred litres per second, and permeability errors can result in a huge increase in flow rate.

## Case history

A shaft 14 m by 8 m was to be constructed to 9 m depth within a cofferdam through a beach deposit of coarse sands and gravels. Permeability was inferred from PSD curves – a $D_{10}$ of approximately 0.5 mm gave a k of $3 \times 10^{-3}$ m/s using Hazen's formula (Equation 5.1). The depth of the gravel aquifer was not proven; boreholes to 20 m bgl did not reach any underlying stratum. The sea was only a few hundred metres away and initial groundwater levels were tidal, up to about 1 m bgl. A system of eight deepwells with a total capacity of approximately 200 l/s was installed. Pumping began at full capacity but lowered the water level by only 1 m. The capacity of the system was roughly doubled by installing another eight wells, which increased the flow rate to 340 l/s; drawdown increased by only 1.5 m. A wellpoint system was also installed inside the cofferdam, but the increase in drawdown was negligible. The dewatering system was now on a very large scale: the wells were at 4 to 5 m spacings and could not be installed much closer, the discharge pipe was 450 mm diameter and a 600 kVA generator was needed to power the system. The system was achieving only 2.5 m drawdown compared with the target of 8 m. Instead of continuing to uprate the dewatering system to achieve an estimated flow rate of nearly 2000 l/s, the dewatering was abandoned and the shaft was excavated and concreted underwater.

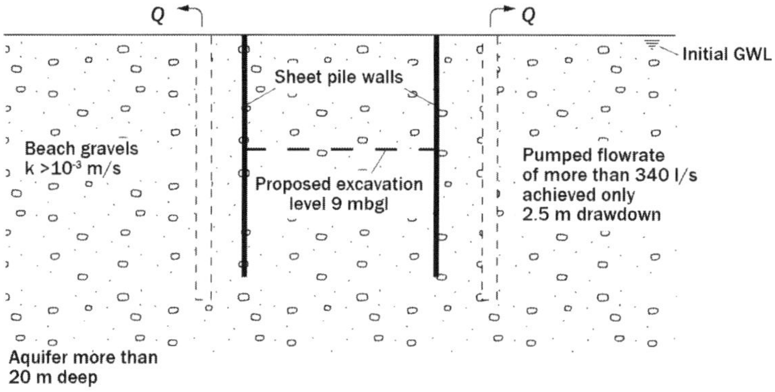


*Figure 7.3 Deepwell system around sheet-piled cofferdam*

## Comment

This is an extreme example of very high flow rates. The problem at this site was particularly acute because of the combination of high permeability, large aquifer thickness and the presence of a nearby recharge source (the sea). The conceptual model at design stage and a permeability sensitivity analysis (Box 6.1) should have revealed the potential for excessive flow rates at design stage. A pumping test would have clarified matters so that underwater construction could have been considered at that stage.

# Case study 7.3 Pore water pressure control in very low permeability soils

## Background

**See also**
2.2.2 Vacuum wellpoints
2.2.3 Vacuum ejector wells

In fine-grained soils such as silts, each well affects such a limited area that individual wells may have to be so closely spaced that a wellpoint system is impractical. If extensive layers of slightly more permeable sand exist in the soil fabric, wellpoint systems may be more effective.

## Case history

In the 1960s an outlet channel for Derwent Reservoir had to be excavated through very sandy (fine) silt with clay and sand partings. PSD analysis showed up to 50 per cent fine sand with silt graded from coarse to fine. The piezometric level was within 1 m of ground level. Initial attempts to excavate using draglines resulted in mud flows, and groundwater control options were considered.

According to Rowe (1968) *"One opinion held that the silt was too fine to be dewatered by any known method. However, an inspection of those parts of the open cut which had not flowed revealed fine layers of sand in the silt ... It also provided ready-made drainage blankets once pore water pressures could be lowered by vacuum wellpoints."* Vacuum wellpoints at 1.2 m centres successfully stabilised the excavation. *"Since the water extraction was achieved via the natural sand layers, once these had been pierced by a representative number of wellpoints, it is likely that a spacing wider than 1.2 m could have been adopted ... therefore the influence of the soil structure can be of paramount importance."*

Cashman (1971) described site conditions and the dramatic improvement in stability following pore water pressure control: *"The first length of the open excavation for the outlet channel was basically waterlogged silt. Soupy silt would be an apt description, though this is not included in standard soil mechanics terminology ... a trial was carried out using wellpoints to test the effectiveness of the technique in the silt. Whereas before the wellpointing it was necessary to wear thigh boots, within a few days after test pumping in that area it was quite possible to exchange them for shoes. The successful draining ... was due mainly, in my view, to the presence of a number of layers of fine sand. These facilitated drainage. It also emphasises that studying the grading envelopes alone may lead one to take a pessimistic view of the feasibility of water lowering. The soil structure itself should also be considered."*

## Comment

A vacuum wellpoint system (see Section 2.2.2) was used successfully, despite the general view that the silt was too fine for such a system. It was adopted because the designer had identified the presence of permeable fabric in the silt. In fine-grained soils fabric can dominate soil drainage (as discussed by Rowe, 1972), so site investigations should be specified to obtain and accurately describe the structure and fabric of high quality soil samples. If the excavation had been carried out in recent years, the use of vacuum ejector wells (Section 2.2.3) might also have been considered.

# Case study 7.4 Effect of low permeability layer

## Background

**See also**

2.1.2 Sump pumping

2.1.9 Sand drains

Soil structure and fabric in the form of low permeability layers may influence groundwater control schemes. Figure 7.4 shows a common situation where, even if an area is generally dewatered, a low permeability layer can leave some residual seepage, known as overbleed.

## Case history

A pumping station was to be constructed in an excavation with battered sides and a wellpoint groundwater control system. Problems occurred with overbleed seepage when a thin stratum of clay was exposed in the face of the batter. Even though the wellpoints had lowered the general water level, some residual water was trapped, or 'perched', above the clay layer and seeped into the excavation. This overbleed caused localised instability of the batter, and work was delayed while a trench drain and sumps were installed as an emergency measure to control the seepage.

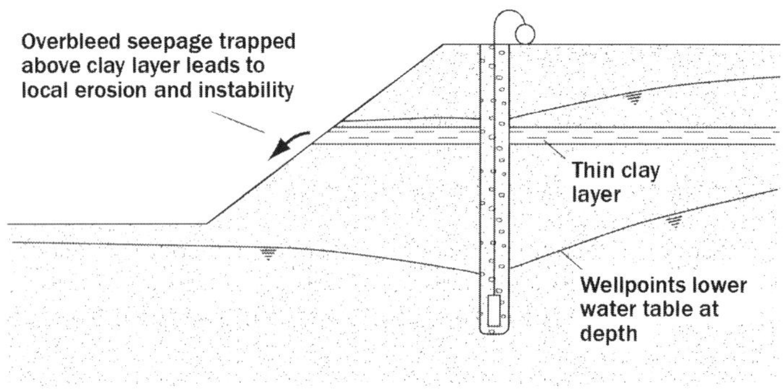


*Figure 7.4 Overbleed seepage*

## Comment

Delay could have been avoided if the conceptual model had identified the clay layer and hence the risk of overbleed seepage. The overbleed could then have been dealt with either by installing the trench drain (Section 2.1.2) as soon as the clay layer was encountered, or by jetting in some sand drains to link the sand above and below the clay layer, draining the perched water (Section 2.1.9).

## Case study 7.5 Instability because of overbleed

### Background

Overbleed seepage can often be easily dealt with in battered excavations where there is room to work, but in small enclosed excavations even small amounts of seepage can cause problems.

### Case history

A shaft 8 m in diameter was to be constructed by underpinning to 10 m depth through 8 m of sandy gravel overlying clay. Deepwells were to be used to lower water levels from 4.5 m bgl to as close to the top of the clay as possible. The design recognised that some residual overbleed seepage would remain over the clay. The sandy gravel was expected to be stable under modest seepage, and it was planned to deal with the overbleed by sump pumping from within the shaft. The system of eight wells lowered the water level to 1.5 m above the clay, but sump pumping led to instability in the shaft face just above the clay and work had to be halted. The problem seemed to be that, despite the overbleed flow being only 2.5 l/s, the soil just above the clay was a silty sand and not a gravel. Silty sands can be very unstable when overbleed occurs and so no significant seepage could be tolerated at the sand–clay interface. This meant a sheet-pile cut-off wall had to be constructed around the shaft to exclude groundwater and allow the shaft to be completed.

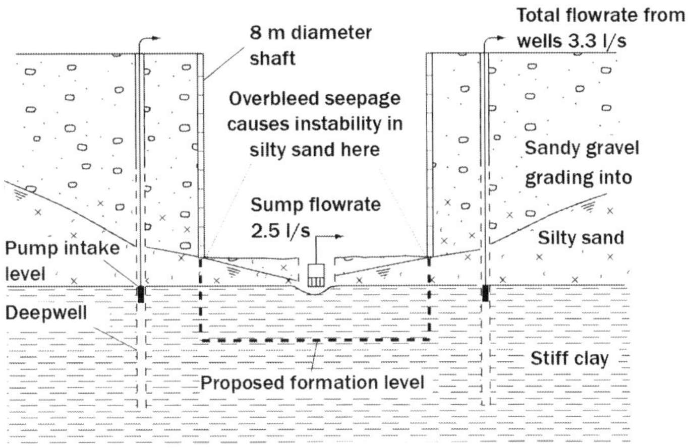


*Figure 7.5 Instability due to overbleed*

### Comment

The presence of the clay stratum above excavation formation level meant that overbleed seepage on the upper surface of the clay was inevitable if pumped well methods were used. If the potential instability of the silty sand layer had been recognised in the conceptual model, alternative construction methods, perhaps such as a ring of closely spaced ejector wells to reduce overbleed seepage, or groundwater exclusion using a cut-off wall, could have been considered at an early stage.

# Case study 7.6 Effect of high permeability shoestring lens (after Preene and Powrie, 1994)

## Background

**See also**
6.1.2 Groundwater flow

Section 6.1.2 considered the need to identify potential aquifer boundary conditions, such as sources of groundwater recharge, when developing the conceptual model. Permeable gravel lenses or 'shoestrings', which may be present in alluvial or fluvio-glacial deposits following old buried stream beds, can be a problem, and very difficult to detect in borehole investigations.

## Case history

A shaft 4 m in diameter was to be constructed to 8 m depth through silty fine to medium sand of fluvio-glacial origin. Based on an anticipated permeability of 1 to 3 $\times$ $10^{-5}$ m/s, equivalent well analysis (Section 6.2.1) predicted a flow rate of 1 to 2 l/s for the required drawdown of 3.5 m. A system of five ejector wells was installed and pumped but achieved only 1.3 m drawdown in the centre of the shaft for 1.4 l/s flow. During excavation one side of the shaft was dry and stable, but seepage occurred on the other side leading to instability and running sand conditions. Mean well yields on the 'wet' side of the shaft were higher than on the 'dry' side. Additional ejector wells were installed, concentrating on the wet side, and eventually the number of ejector wells was increased from 5 to 22. Three of the extra wells encountered a water-bearing lens or shoestring of coarse gravel a few metres from the wet side of the shaft. The wet side of the shaft dried up, allowing the works to be completed: total flow rate was 3.7 l/s from the ejectors.

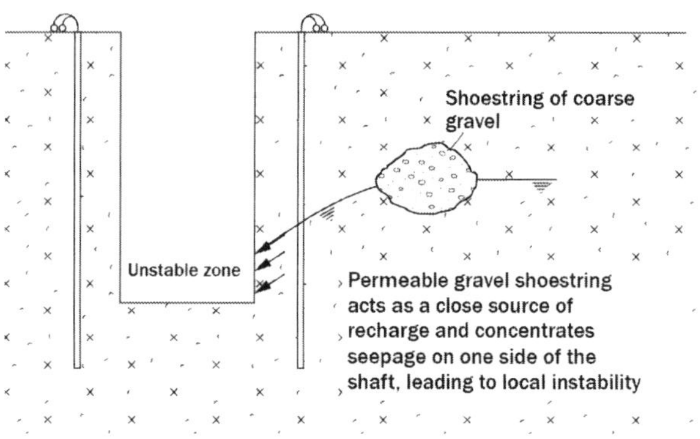


*Figure 7.6 Instability due to seepage from shoestring lens*

## Comment

The gravel shoestring probably acted as a conduit drawing water toward the dewatering system, forming a very localised source of recharge. The shaft was not stabilised until some wells tapped directly into the shoestring. The thin, linear nature of the shoestring makes detection by ground investigation largely a matter of chance. The problem was so localised that the shaft could probably have been completed using the original system if the gravel shoestring had been just a few metres further away. If there is an indication that such features may be present (eg in alluvial or fluvio-glacial soils), an appropriate conceptual model should allow for them.

# Case study 7.7 Wellpoint and ejector well systems used in combination

## Background

See also
1.2.6 Objectives of groundwater control

Boundary conditions identified in the conceptual model can influence the selection of groundwater control methods, especially if there is more than one potential aquifer or a low permeability layer. It can be difficult for one pumping technique to deal with both high and low permeability soils; in some cases it may be necessary to use a combination of pumping techniques.

## Case history

An underbridge was to be constructed by jacking a concrete box beneath an existing railway embankment. Excavation within the box was to be below initial groundwater levels through coarse terrace gravels over less permeable silty sands of the Bracklesham Beds. The conceptual model predicted significant inflows from the gravels, which meant that pumping would be required to prevent the excavation flooding, but also that much smaller flow rates, if pumped from the silty sand, would control pore water pressures and prevent quicksand conditions. A single groundwater control technique was unlikely to be able to deal with both strata at once, so the solution adopted was to use two in combination. A wellpoint system was used to lower water levels in the gravel and an ejector well system was used to reduce pore water pressures in the silty sand. An additional complication was that wells could only be drilled from either side of the railway, so several ejector wells were installed at an angle to form a 'fan' of wells beneath the embankment.

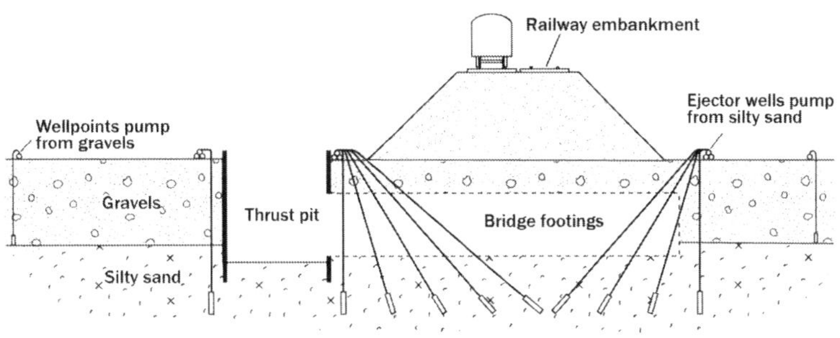


*Figure 7.7 Wellpoint and ejector systems in combination*

## Comment

Because of the difference in behaviour (see Section 1.2.6) of coarse-grained soils (eg gravels), where the pore water can drain freely, and fine-grained soils (eg silty sands), which drain less freely (but where pore water pressure reductions can give dramatic improvements in stability), each soil needs to be dealt with in a different way. In the coarse-grained soil wellpoints were intended to pump large flow rates, and in the fine-grained soil the ejector wells were intended to control pore water pressures.

# Case study 7.8 Assessment of settlement risk at feasibility stage

## Background

See also
3.2 CDM Regulations
Chapter 4 Environmental matters
6.6 Settlement

External factors may affect the application of groundwater control techniques. Settlement analysis has been described in Section 6.6, and Chapter 4 has described some of the environmental effects of pumping. As the conceptual model is developed, potential risks and hazards should be identified and assessed in accordance with CDM Regulations (Section 3.2).

## Case history

A structure 9 m deep was to be constructed approximately 20 m from an existing deep shaft (which had been built 30 years previously using groundwater control techniques). Ground conditions consisted of 10 m of soft silty clay over a variable succession of interlayered alluvial sand and clay deposits underlain by very stiff clay at a depth of 20 m. Initial groundwater levels were close to ground level. Groundwater control by either ejector wells or deepwells appeared to be feasible, but effective stress calculations (Section 6.6) indicated the potential for settlements of 100 mm to 150 mm adjacent to the structure, decreasing further away. On a greenfield site these settlements might not have been critical (construction of the existing pumping station had probably generated similar settlements). However, the site was now crossed by a sewer, which would settle with the surrounding ground. This sewer was connected into the existing shaft, which was founded on piles bearing on the very stiff clay, and so would settle much less than the sewer. Groundwater lowering might induce differential settlements in excess of 50 mm where the sewer met the existing structure. There would have been a significant risk of the sewer rupturing at that point, with disastrous consequences for the sewerage system in the surrounding area. As a result, the contractor did not attempt any dewatering, but used the more expensive method of constructing a complete physical cut-off wall around the new structure and monitoring groundwater levels to check that no inadvertent groundwater lowering occurred from sump pumping from within the works. The extra cost was justified by the reduced risk of damage to the sewer.

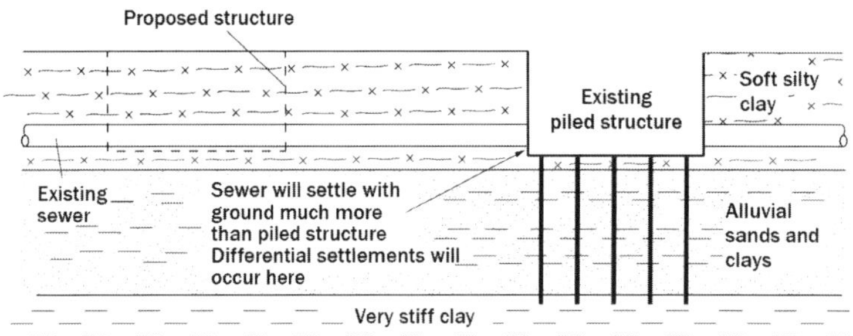


*Figure 7.8 Settlement risk to sewer*

## Comment

This case study is interesting in two ways. Firstly, pumping had previously been carried out at the site and no settlement damage had occurred, because the vulnerable infrastructure (the sewer) had not then been constructed. Secondly, the major cause of concern was not the absolute settlements but, as is often the case, the differential settlements where the sewer met the existing structure.

## Case study 7.9 Groundwater control in an urban area (after Cashman, 1987 and 1994a)

### Background

In urban areas, groundwater control may be complicated by the presence of nearby structures and the problem of disposing of the discharge water. Human factors can also play a part.

### Case history

In the 1980s a new bank headquarters was constructed in the centre of Cairo, Egypt. Given the proximity of the surrounding buildings, drawdowns outside the site had to be controlled and monitored to minimise settlement risks. Wellpoints inside a sheet-piled cofferdam were pumped to control pore water pressures within the excavation and the resulting discharge (28 l/s to 42 l/s) was disposed of via recharge wells outside the cofferdam. By monitoring piezometers, the pumping rates were adjusted so that external water levels did not move outside prescribed limits. Without such a recharge system, it is unlikely that the Cairo authorities would have allowed the project to proceed. Use of recharge had an additional benefit in that it avoided having to discharge to the Cairo sewer system, which was heavily overloaded and might not have coped with the extra flow. Geotechnical reasons (control of settlements) for applying recharge may have been secondary to practical considerations (disposal of discharge flow).

This project also highlighted the human element in any groundwater control system. Cashman (1987) recalled that *"our field supervisor had really not a lot of faith in recharge. He tapped into the Cairo sewer system with a hidden discharge pipe and most of the water of the discharge system was going there. Unfortunately ... between Christmas and New Year, one of the Cairo main pumping stations broke down – that does happen quite frequently there – and everything flooded back. As a result the chairman of the main constructor's company received a telephone call personally from the mayor of Cairo municipality demanding his personal presence on site immediately. He was told that if such a thing ever happened again, he, the chairman, would immediately be put in jail."* This was a pretty strong incentive to keep the system going.

The project is described in more detail by Troughton (1987).

## Case study 7.10 Large dewatering scheme using perimeter deepwells

### Background

Even large excavations can often be dewatered with a perimeter well scheme. The use of numerical modelling to support the design of a dewatering system is appropriate where the drawdown is time dependent and also where there is concern about potential settlement impact on neighbouring structures. Initial flows to achieve drawdown may be appreciably greater than the steady state flows required to maintain drawdown.

### Case history

A new terminal was required at an international airport in the Middle East, Roberts *et al* (2009). Lack of space dictated that the new terminal was to be constructed underground below the aircraft taxiway and apron. Ground conditions comprised Aeolian dune sand overlying weakly cemented sandstone with standing groundwater at approximately 3 m bgl. The excavation involved the removal of 13 million $m^3$ of soil over an area of 420 000 $m^2$ (450 m by 700 m) to a depth of 25 m bgl. Excavation side support was provided by 3.5 km of diaphragm wall (with temporary anchor support), which was to be incorporated into the permanent works structure. The diaphragm wall was not designed to take external groundwater pressure in the temporary condition. An initial steady state analysis based on data from two pumping tests indicated a requirement for a perimeter ring of wells at 25 m spacing with a total abstraction flow of 194 l/s. Access was available to the excavation in phases as areas of the existing airport were vacated and cleared of services. A three-dimensional time dependent groundwater numerical model was developed and calibrated against the pumping test data and the first phases of pumping. The model was used to predict the development of the drawdown and to check this against the programme for excavation and access for subsequent well installation and commissioning. The model also provided an assessment of the drawdowns generated below adjacent structures. In the event maximum discharge flows peaked at 290 l/s but reduced to about 100 l/s approximately 18 months after pumping commenced. The dewatering scheme was used to maintain the drawdown for a further two years while the terminal was completed.

### Comment

The dewatering of this large excavation was carried out effectively with a perimeter well array although temporary internal wells were used as part of the phased access to the works. The large areas involved meant that the drawdown was time dependent and flows reduced appreciably once the drawdown was established. This reduction in flow was accompanied by an increase in the distance of influence of the dewatering scheme. The contour plot outputs from the numerical groundwater model provided a useful means of communicating the development and extent of the drawdown cone to the rest of the project team. In the event, as expected, the ground was sufficiently dense and stiff, that no significant settlement impact on adjacent structures was recorded by the monitoring.

# Case study 7.11 Artificial recharge used to control settlement risk

## Background

Older structures were sometimes built with poor foundations, which can lead to challenges when new infrastructure is to be built nearby. While the dewatering of dense granular soils may be lead to minimal surface settlements the underdrainage of soft alluvial soils above can lead to significant consolidation and corresponding surface settlement.

## Case history

The construction of a new high speed railway line required a tunnel section beneath the Thames in the east of London, Roberts and Holmes (2010). Deep excavations were required for the two portal structures, which included the tunnel launch and reception chambers as well as cut-and-cover and open cut sections. The portal structures were located on the river flood plain where ground conditions comprised up to 10 m of soft alluvial clays and silts overlying terrace gravels and upper chalk. Standing groundwater level was at ground level and the maximum excavation level reached down to the top of the chalk at approximately 18 m bgl. Excavation side support was provided by a propped diaphragm wall, which toed in to the chalk. Adjacent to the north portal was a petroleum products tank farm, which was assessed as having a high sensitivity to settlement. Numerical modelling showed that in the absence of mitigation measures drawdown in the terrace gravels below the adjacent tank farm was likely to be up to 4 m. This would have led to the underdrainage and consolidation of the alluvial soils above generating unacceptable surface settlements. In order to mitigate this risk a system of recharge wells was installed around the tank farm to maintain the groundwater levels in the terrace gravels. The design of the recharge scheme was based on the numerical model, which was calibrated using the results from an abstraction/recharge pumping test and a programme of permeability testing at different depths. A study had concluded that the alluvium could sustain a change of effective stress of 5 kPa (equivalent to 0.5 m drawdown) over the 60 week period of the dewatering without measurable change in ground surface level. The aim of the recharge scheme was then to limit drawdown to 0.5 m below the ambient tidal cyclic groundwater level. Amber and red trigger levels were set at 0.75 m and 1.25 m drawdown respectively on the basis that excess drawdowns were acceptable in the short term. Trigger levels were also set for settlement monitoring but it was recognised that groundwater levels, being a lead indicator, were key to controlling ground movements. In the event groundwater levels were controlled satisfactorily with only a single breach of an amber trigger level due to insufficient attention being paid to recharge well cleaning, which was swiftly remedied. No significant surface settlement was recorded at the tank farm whereas settlements of up to 100 mm were recorded in other areas where there were no neighbouring structures and no recharge.

## Comment

In this case the settlement risk was identified and thoroughly investigated prior to the start of the works. As a result an effective mitigation measure comprising an artificial recharge scheme was implemented. The inflow and external drawdown under a partial cut-off is controlled primarily by the vertical permeability whereas most borehole permeability tests and conventional single well pumping tests measure the horizontal permeability. Data obtained from an abstraction/recharge test and permeability testing combined with a numerical model proved an effective strategy for developing the scheme. The modelling showed that extending the cut-off deeper in to the chalk was advisable to curtail excessive recirculation of the recharged groundwater and the diaphragm wall was extending accordingly.

## Case study 7.12 Groundwater control in multiple aquifers

### *Background*

The way in which multiple aquifers interact can be difficult to assess without local experience. Sometimes the differences in aquifer response and hydraulic connection can be exploited to advantage.

### *Case history*

A new underground railway required a 30 m diameter by 44 m depth temporary drive shaft in East London. The shaft was constructed within a 55 m deep diaphragm wall and was to be used for ventilation of the permanent works. Stratification at the site is typical of East London comprising:

- made ground/alluvium — to 12 m depth
- terrace gravels — 12 m to 14 m depth
- London Clay — 14 m to 44 m depth
- Lambeth Group channel sands — 44 m to 48 m depth
- Lambeth Group clays — 48 m to 52 m depth
- Thanet Sand — 52 m to 78 m depth
- chalk — below 78 m depth.

Standing groundwater level was at approximately 10 m depth. The terrace gravels (upper aquifer) were cut-off by the diaphragm wall and did not require dewatering. Pressure relief was required in the Lambeth Group channel sands (intermediate aquifer) and in the Thanet Sand (lower aquifer). The Lambeth Group channel sands have minimal hydraulic connection to the Thanet Sand below due to the low permeability clay horizons that make up much of the Lambeth Group stratification. Experience has shown that it can be preferable to underdrain the Thanet Sand by pumping from the underlying chalk. The downside of this approach is the higher flow rates derived from the chalk. Evidence from past projects shows that the hydraulic connection between the chalk and Thanet Sand is often constrained by an increase in silt content towards the base of the Thanet Sand. The strategy adopted for dewatering of the shaft comprised:

- **Lambeth Group channel sands:** internal passive relief wells.
- **Thanet Sand:** partial drawdown achieved using external chalk wells to underdrain the Thanet Sand. This was supplemented by external Thanet Sand wells installed before excavation. Internal Thanet Sand wells were also installed in the partially excavated shaft.

The chalk wells plus the external and internal Thanet Sand wells were pumped on whilst the excavation was bottomed out and the base slab cast. Once the slab gained sufficient strength the weight of the diaphragm wall could be mobilised to resist the hydrostatic uplift pressures and the external chalk and Thanet Sand wells were shut down. The modest drawdown required during the three year tunnelling period was maintained by passive relief into the shaft from the Lambeth Group and internal Thanet Sand wells. The completed ventilation shaft had sufficient weight to resist the hydrostatic loads allowing the relief wells to be sealed.

### *Comment*

The Lambeth Group channel sands were hydraulically isolated from the Thanet Sand below and cut-off by the diaphragm wall. This was confirmed by test pumping one of the relief wells before excavation. This testing proved the integrity of the base plug which also minimised the drawdown required in the Thanet Sand. The full drawdown required for shaft excavation and casting of the base slab was only needed for a comparatively short time. Pumping of the higher flows derived from the chalk was avoided during the three year tunnelling programme by using the internal Thanet Sand relief wells.

# Case study 7.13 Basement excavation in gravels

## Background

Many cities are built around rivers and estuaries and increasingly urban development involves the exploitation of underground space. This space may be used for transport infrastructure, car parking or utility services. For large area developments in thick aquifers provision of an effective temporary cut-off could require very deep vertical walls or horizontal grout cut-offs, which may be prohibitively costly. A partial cut-off with dewatering can represent a cost-effective alternative solution.

## Case history

A new city centre development required an excavation for a double basement in water bearing fluvio glacial gravels adjacent to a tidal river, see Long *et al* (2007). A partial cut-off was provided by an anchored sheet-pile cofferdam. The clutches of the sheet-piles were to be welded up to provide the permanent basement wall. A pumping test and local experience of dewatering open cut excavations (with no cut-off) indicated that the gravels were of potentially high horizontal permeability ($>10^{-3}$ m/s) implying very heavy inflows approaching the upper economic limit for a pumped dewatering scheme (see Figure 1.10) given the amount of drawdown required (7 m below high tide level in the river). The works proceeded with provision for a deep well scheme with a total flow capacity of up to 500 l/s. In the event abstraction flows peaked at 280 l/s and back analysis suggested that the permeability of the gravels around the toe of the sheet piles was probably of the order of $5 \times 10^{-4}$ m/s – appreciably lower than the permeability of the superficial gravels. Subsequent experience on other sites in the city, including more detailed site investigation studies, suggest that the sand content of the gravels increases at about the toe level of the sheet piles. This was thought to be the depth to which reworking of the glacial gravels may have occurred due to fluvial (river flow) action resulting in removal of the sand fraction from the shallower gravels.

## Comment

There is no simple way to determine the permeability profile in a relatively permeable aquifer. A pumping test combined with borehole logs, particle size distribution testing of representative samples at a range of depths and possibly *in situ* variable head testing can give some guidance. A modest reduction in permeability with depth and anisotropic conditions (vertical permeability less than the horizontal permeability) are commonly observed characteristics of even relatively uniform soils. These conditions can have a significant impact on seepage flows below partial cut-offs.

## Case study 7.14 River wells and tunnel wellpoints

### Background

A good understanding of the ground conditions and an accurate conceptual model are vital ingredients when planning a dewatering scheme. The most important source of recharge may not be the most immediate or visible source.

### Case history

A cross passage and central sump was to be built between two completed tunnel boring machine (TBM) driven tunnels 30 m below a tidal river in uniform fine sand. A previous attempt to construct the cross passage using 3 bar compressed air pressure was unsuccessful and resulted in the damage and flooding of one of the TBM tunnels. Artificial ground freezing was considered as an option to progress the works but required access to both TBM tunnels, which would have involved a significant delay to the programme. A review of the ground profile revealed the presence of a 4 m thick clay horizon below river bed level. Experience suggested that this provided an effective hydraulic barrier between the river and the fine sand aquifer below. This allowed the development of a dewatering strategy comprising the following elements:

- 8 no. deepwells drilled into the river bed and operated by submersible pumps.
- 10 no. wellpoints up to 10 m length installed from the dry TBM tunnel into the sand aquifer.
- Compressed air working at <1 bar.

The wells were terminated at river bed level in order not to provide a hazard to navigation. Pumps were installed on flexible riser pipes with a bolted top flange by divers. To minimise the risk of an interruption to pumping half of the pumps were powered from a generator on a barge anchored above the wells with the remaining pumps powered from generators on shore. Monitoring was carried out using submersible flow meters and vibrating wire transducers. The tunnel wellpoints were installed through a stuffing box bolted to the tunnel lining in order to control ingress of sand and groundwater during installation. Once installed the wellpoints were connected to a headermain and vacuum wellpoint pump located in the TBM tunnel outside the compressed air lock. In the event, with the deepwells and tunnel wellpoints in operation, the cross passage was constructed in free air. Compressed air working was used for construction of the sump. On completion the river wells were extended up to above river level temporarily so that they could be backfilled and sealed before being cut-off at 1 m below river bed level.

### Comment

The tidal river was a very obvious and immediate potential source of recharge. However good knowledge of the ground profile combined with relevant local experience led to the development of a conceptual model, which showed the river was not an important source of recharge to the aquifer of concern. The success of the scheme confirmed the validity of this important assumption. The use of sub-aqueous wells is not unprecedented – their application is described for a large project by Doran *et al* (1995), and Biggart and Sternath (1996). Installing wellpoints from tunnels can be an effective way to overcome surface access restrictions but careful consideration of the risk and challenges is required, together with experienced site staff, in order to achieve successful installation.

## 7.4 CONCLUSION

See also
Figure 6.1 Design

This chapter uses case studies to highlight some lessons in the design and implementation of groundwater control systems. The most important lesson is that, to avoid delays and unnecessary costs, groundwater control requirements should be planned for from the start of a project through to its end (see Figure 6.1). Experience suggests that successful groundwater control projects involve the following stages, whether carried out by one or several organisations, depending on the contractual framework for the project:

1. Assessment of potential groundwater problems during the design of permanent and temporary works, including environmental questions, where possible selecting appropriate techniques at an early stage.
2. Execution of a site investigation designed to provide the information needed for groundwater control systems.
3. Consultation with the appropriate environmental regulator or authority to obtain the necessary consents.
4. Use of design methods, which concentrate on getting the conceptual model right and selecting appropriate permeability values.
5. Methods of analysis and calculations which use sensitivity or parametric analyses to assess the effect of variations in permeability or boundary conditions. It is not realistic to expect a set of unique answers from calculations, and it is better to predict a range of values of, say, flow rate.
6. Design and specification of a flexible system, which can be easily modified to meet the range of analytical results (eg flow rate, time to achieve drawdown).
7. Supervision of the installation of the system to make sure it is carried out correctly.
8. Monitoring and analysis of the performance of the system at start up and during the initial drawdown period, in order to facilitate a prompt response if modifications are necessary.
9. Maintenance and monitoring during the operational period.
10. Review of the groundwater control aspects on completion of the project and dissemination of data.

# References

AL-DHAHIR, Z A, KENNARD, M F and MORGENSTERN, N R (1969) "Observations on pore pressures beneath the ash lagoon embankments at Fiddler's Ferry power station". In: *Proc of the conf on in situ investigations in soils and rocks, British Geotechnical Society, London*, 13–15 May 1969, Institution of Civil Engineers, London

ANDERSON, M P and WOESSNER, W W (2015) *Applied groundwater modelling, second edition*, Academic Press Inc, New York, USA (ISBN: 978-0-12058-103-0)

ASPINWALL AND COMPANY (1994) *A framework for assessing the impact of contaminated land on groundwater and surface water*, Contaminated Land Report (CLR) 1, Department of the Environment, London

ATTEWELL, P B (1995) *Tunnelling contracts and site investigation*, CRC Press, Boca Raton, USA (ISBN: 978-0-41919-140-7)

BDA (2008) *Guidance for safe intrusive activities on contaminated or potentially contaminated land*, British Drilling Association, Nottinghamshire, UK

BDA (2015) *Health and safety manual for land drilling 2015: a code of safe drilling practice*, British Drilling Association, Nottinghamshire, UK

BEAR, J (1979) *Hydraulics of groundwater*, McGraw-Hill International, New York (ISBN: 978-0-07004-170-7)

BELL, A L (ed) (1993) *Grouting in the ground*, Thomas Telford Publishing, London (ISBN: 978-0-72771-928-7)

BELL, F G and MITCHELL, J K (1986) "Control of groundwater by exclusion". In: J C Cripps, F G Bell and M G Culshaw (eds) *Groundwater in engineering geology*, Geological Society Special Publication 3, Geological Society, London, pp 429–443

BEVAN, M A, POWRIE, W and ROBERTS, T O L (2010) "Influence of large-scale inhomogeneities on a construction dewatering system in chalk" *Géotechnique*, vol 60, 8, Institution of Civil Engineers, UK, pp635–649

BIGGART, A R and STERNATH, R (1996) "Storebaelt eastern railway tunnel: construction" *Civil Engineering*, vol 114, Storebaelt Eastern Railway Tunnel, Supplement, Institution of Civil Engineers, London, pp 20–39

BOAK, R, BELLIS, L, LOW, R, MITCHELL, R, HAYES, P, McKELVEY, P and NEALE, S (2007) *Hydrogeological impact appraisal for dewatering abstractions*, Science Report SC040020/SR1, Environment Agency, Bristol (ISBN: 978-1-84432-673-0)

BOCK, M, and MARKUSSEN, L M (2007) "Dewatering control in central Copenhagen". In: *Proc of the 14th European Conference on Soil Mechanics and Geotechnical Engineering, Madrid, Spain*, 24–27 September 2007, V Cuéllar, E Dapena, E Alonso, J M Echave, A Gens, J L De Justo, C Oteo, J M Rodríguez-Ortiz, C Sagaseta, P Sola, A Soriano (eds) *Geotechynical engineer in urban environments*, vol 2, IOS Press, Amsterdam, The Netherlands, (ISBN: 978-9-05966-055-7), pp 715–720

BOLTON, M D (1991) *A guide to soil mechanics, third edition*, M D and K Bolton, Cambridge (ISBN: 978-0-95227-650-0)

BOND, A (ed) (1994) *Validation and use of geotechnical software*, Association of Geotechnical and Geoenvironmental Specialists (AGS), Beckenham, Kent (ISBN: 978-0-95192-714-4)

BRAND, E W and PREMCHITT, J (1982) "Response characteristics of cylindrical piezometers" *Géotechnique*, vol 32, 3, Institution of Civil Engineers, London, pp 203–216

BRANDON, T W (ed) (1986) "Groundwater, occurrence, development and protection" *Water practice manual No. 5*, Chartered Institution of Water and Environmental Management, London (ISBN: 978-0-90142-714-4)

BRASSINGTON, R (2006) *Field hydrogeology, thid edition*, Wiley-Blackwell, London (ISBN: 978-0-470-01828-6)

BRASSINGTON, F C and PREENE, M (2003) "The design, construction and testing of a horizontal wellpoint in a dune sands aquifer as a water source" *Quarterly Journal of Engineering Geology and Hydrogeology*, vol 36, 4, The Geological Society of London, London, pp 355–366

BURLAND, J B and WROTH, C P (1975) "Settlement of buildings and associated damage". In: *Proc British Geotechnical Society conference, Lady Mitchell Hall, Cambridge*, April 1974, A C Meigh (ed) Settlement of structures, Pentech Press, London (ISBN: 978-0-72731-901-2), pp 611–654

CARTER, M (1983) *Geotechnical engineering handbook*, Pentech Press, California, USA (ISBN: 978-0-72730-702-6)

CASAGRANDE, L (1952) *Electro-osmotic stabilization of soils*, Harvard University Press, Boston, USA

CASAGRANDE, L, WADE, N, WAKELY, M and LOUGHNEY, R (1981) "Electro-osmosis projects". In: *Proc of the 10th int conf on soil mechanics and foundation engineering, Stockholm, Sweden, Balkema, Rotterdam*, 15–19 June 1981 (ISBN: 978-9-06191-210-1) pp 607–610

CASHMAN, P M (1971) "Discussion on Derwent Dam" *Proceedings of the Institution of Civil Engineers*, vol 48, 3, Institution of Civil Engineers, London, pp 479–521

CASHMAN, P M (1987) "Discussion". In: *Proc of the 9th conf on soil mechanics and foundations, Dublin, Ireland*, 31 August 3 September 1987, vols 1–3, E T Hanrahan, T L L Orr, T F Widdis (eds) *Groundwater effects in geotechnical engineering*, CRC Press, Boca Raton, USA (ISBN: 978-9-06191-720-4), p 1015

CASHMAN, P M (1994a) "Discussion". In: *Proc of the int conf organised by the Institution of Civil Engineers, London*, 2–3 June 1993, W B Wilkinson (ed) *Groundwater problems in urban area*, Thomas Telford, London (ISBN: 978-0-72771-974-4), pp 93–96

CASHMAN, P M (1994b) "Discussions of Roberts and Preene". In: *Proc of the int conf organised by the Institution of Civil Engineers*, London 2–3 June 1993, W B Wilkinson (ed) *Groundwater problems in urban area*, Thomas Telford, London (ISBN: 978-0-72771-974-4), pp 446–450

CASHMAN, P M and PREENE, M (2012) *Groundwater lowering in construction: a practical guide to dewatering, second edition*, CRC Press, Boca Raton, USA (ISBN: 978-0-41921-110-5)

CEDERGREN, H R (1989) *Seepage, drainage and flow nets, third edition*, Wiley, New York (ISBN: 978-0-47118-053-1)

CHAPMAN, T G (1959) "Groundwater flow to trenches and wellpoints" *Journal of the Institution of Engineers, Australia*, October–November, Institution of Engineers Australia, pp 275–280

CLAYTON, C R I (1995) *The Standard Penetration Test (SPT): methods and use*, R143, CIRIA, London (ISBN: 978-0-86017-419-6). Go to: **www.ciria.org**

CLAYTON, C R I, MATTHEWS, M C and SIMONS, N E (1995) *Site investigation: a handbook for engineers, second edition*, Wiley-Blackwell, London (ISBN: 978-0-63202-908-2)

CLOUGH, G W and O'ROURKE, T D (1990) "Construction induced movements of in-situ walls". In: *Proc of ASCE conference, Cornell University, Ithaca, New York, USA*, 18–21 June 1990, P Lambe, L A Hansen (eds) Design and performance of earth retaining structures, American Society of Civil Engineers, New York, USA (ISBN: 978-0-87262-761-1), pp 439–470

COOMBER, D B (1986) "Groundwater control by jet grouting". In: J C Cripps, F G Bell, M G Culshaw (eds) *Groundwater in engineering geology*, Geological Society Special Publication 3, Geological Society, London, pp 445–454

COOPER, H H and JACOB, C E (1946) "A generalized graphical method for evaluating formation constants and summarising well field history" *Transactions of the American Geophysical Union*, vol 27, 4, Wiley Online, pp 526–534

CRIPPS, J C, BELL, F G and CULSHAW, M G (eds) (1986) *Groundwater in engineering geology conference proceedings*, The Geological Society, London (ISBN: 978-0-90331-735-1)

DEFRA (1987) *Guidance on the assessment and redevelopment of contaminated land*, Guidance Note 59/83, Department for the Environment, Food and Rural Affairs, London

DORAN, S R, HARTWELL, D J, ROBERTI, P, KOFOED, N and WARREN, S (1995) "Storebaelt railway tunnel – Denmark: implementation of cross passage ground treatment". In: *Proc of the 11th European conference on soil mechanics and foundation engineering, Copenhagen, Denmark*, 12–16 August 1995

DRISCOLL, R, SCOTT, P and POWELL, J (2008) *EC7 – implications for UK practice*, C641, CIRIA, London (ISBN: 978-0-86017-641-1). Go to: **www.ciria.org**

ENVIRONMENT AGENCY (2012) *Good practice for decommissioning redundant boreholes and wells*, Environment Agency, Bristol, UK. Go to: **http://tinyurl.com/jzybg4x**

ERVIN, M C and MORGAN, J R (2001) "Groundwater control around a large basement" *Canadian Geotechnical Journal*, vol 38, 4, NRC Research Press, Canada, pp 732–740

FETTER, C W (2014) *Applied hydrogeology, fourth edition*, Pearson New International Edition, Pearson Education Limited, Essex (ISBN: 978-1-29202-290-1)

FREEZE, R A and CHERRY, J A (1979) *Groundwater*, Prentice-Hall, Englewood Cliffs, New Jersey, USA (ISBN: 978-0-13365-312-0)

FUSSELL, D R, GODJEN, H, HAYWARD, P, LILIE, R H and MARCO, A (1981) *Revised inland oil spill clean-up manual*, CONCAWE-7/81, CONCAWE, The Hague, the Netherlands

GABA, A R, SIMPSON, B, POWRIE, W and BEADMAN, D R (2003) *Embedded retaining walls – guidance for economic design*, C580, CIRIA, London (ISBN: 978-0-86017-580-3). Go to: **www.ciria.org**

GILBERTSON (2015) *CDM2015 – workplace 'in-use' guidance for designers, second edition*, C756, CIRIA, London (ISBN: 978-0-86017-762-3). Go to: **www.ciria.org**

GODFREY, P S (1996) *Control of risk: a guide to the management of risk from construction*, SP125, CIRIA, London (ISBN: 978-0-86017-441-7). Go to **www.ciria.org**

HANRAHAN, E T, ORR, T L L and WIDDIS, T F (eds) (1987) *Groundwater effects in geotechnical engineering, volume 2: Proceedings of the 9th European conference on soil mechanics and foundation engineering, Dublin, Ireland, 31 August to 3 September 1987*, Taylor and Francis, UK (ISBN: 978-9-06191-722-9)

HARRIS, J S (1995) *Ground freezing in practice*, Thomas Telford Publishing, London (ISBN: 978-0-72771-995-9)

HARRIS, M R, HERBERT, S M and SMITH, M A (1995) *Remedial treatment for contaminated land*, vol I–XII, SP164, CIRIA, London (ISBN: 978-0-86017-408-0). Go to: **www.ciria.org**

HARTWELL, D J and NISBET, R M (1987) "Groundwater problems associated with the construction of large pumping stations". In: *Proc of the 9th conf on soil mechanics and foundations engineering, Dublin, Ireland*, 31 August to 3 September 1987, E T Hanrahan, T L L Orr and T F Widdis (eds) Groundwater effects in geotechnical engineering, CRC Press, Boca Raton, USA (ISBN: 978-9-06191-720-5), pp 691–694

HAUSMANN, M R (1990) *Engineering principles of ground modification*, McGraw-Hill, New York, USA (ISBN: 978-0-07027-279-8)

HOLDEN, J M W, JONES, M A, FERNANDO, M-W and WHITE, C (1998) *Hydraulic measures for the treatment and control of groundwater pollution*, R186, CIRIA, London (ISBN: 978-0-86017-499-8). Go to: **www.ciria.org**

HOULSBY, A C (1976) "Routine interpretation of the Lugeon water test" *Quarterly Journal of Engineering Geology and Hydrogeology*, vol 9, The Geological Society, London, pp 803–814

HOWSAM, P (ed) (1990) *Microbiology in civil engineering: international proceedings of the Federation of European Microbiologial Societies symposium held at Cranfield Institute of Technology, UK*, 3–5 September 1990, E & F Spon, UK (ISBN: 0-41916-730-7), FEMS Symposium), Spon, London

HOWSAM, P, MISSTEAR, B and JONES, C (1995) *Monitoring, maintenance and rehabilitation of water supply boreholes*, R137, CIRIA, London (ISBN: 978-0-86017-417-2). Go to: **www.ciria.org**

HSE (2006) *Health and safety in construction, third edition*, HSG150, Health and Safety Executive, London (ISBN: 978-0-71766-182-2). Go to: **http://tinyurl.com/q5fdm6n**

HSE (2015) *Managing health and safety in construction: Construction (Design and Management Regulations) 2015. Guidance on Regulations*, L153, Health and Safety Executive, London (ISBN: 978-0-71766-626-3). Go to: **www.hse.gov.uk/pubns/priced/l153.pdf**

ICE (1982) "Third Géotechnique symposium in print: vertical drains" *Géotechnique*, vol 31, 1, Thomas Telford Ltd, London, pp 1–90

ICE (1991) *Inadequate site investigation*, Thomas Telford Ltd, London (ISBN: 0-72771-645-X)

IRVINE, D J and SMITH, R H J (2001) *Trenching practice, revised edition*, R97, CIRIA, London (ISBN: 978-0-86017-192-8). Go to: **www.ciria.org**

JEFFERIS, S A (1993) "In-ground barriers". In: T Cairney (ed) *Contaminated land – problems and solutions, second edition*, CRC Press, Boca Raton, USA (ISBN: 978-0-41923-090-8), pp 111–140

KENNEDY, R A, LLOYD, J W and HOWLEY, J A (1988) "Aspects of geotextile-wrapped well screen design – an experimental investigation" *Quarterly Journal of Engineering Geology and Hydrogeology*, vol 21, 2, The Geological Society, London, pp 137–145

KENNEY, T C, CHAHAL, R, CHIU, E, OFOEGBU, G I and UME, C A (1985) "Controlling constriction size of granular filters" *Canadian Geotechnical Journal*, vol 22, NRC Research Press, Canada, pp 32–43

KING, J M (1984) "Computing drawdown distributions using microcomputers" *Groundwater*, vol 22, 6, Wiley Online, UK, pp 780–784

KNIGHT, D J, SMITH, G L and SUTTON, J S (1996) "Sizewell B foundation dewatering – system design, construction and performance monitoring" *Géotechnique*, vol 46, 3, Institution of Civil Engineers, London, pp 473–490

KOFOED, N and DORAN, S R (1995) "Storebaelt tunnel: groundwater modelling for cross passages". In: *Proc 11th European conference on soil mechanics and foundation engineering, Copenhagen, Denmark*, 12–16 August 1995

KRUSEMAN, G P and DE RIDDER, N A (1990) *Analysis and evaluation of pumping test data, second edition*, Publication 47, International Institute for Land Reclamation and Improvement, Wageningen, The Netherlands (ISBN: 978-9-07075-420-4)

LEIPER, Q, ROBERTS, T O L, and RUSSELL, D (2000) "Geotechnical engineering for the Medway tunnel and approaches" *Proceedings of the ICE – Transportation*, vol 141, 1, Institution of Civil Engineers, London, pp 35–42

LONG, M, ROBERTS, T O L and CREED, M (2007) "Soil characterisation and construction of deep basements in high permeability gravels in Cork, Ireland". In: *Proc of the 14th European conference on soil mechanics and geotechnical engineering, Madrid, Spain*, 24–27 September 2007, V Cuéllar, E Dapena, E Alonso, J M Echave, A Gens, J L De Justo, C Oteo, J M Rodríguez-Ortiz, C Sagaseta, P Sola, A Soriano (eds) *Geotechnical engineer in urban environments*, vol 2, IOS Press, Amsterdam, The Netherlands, (ISBN: 978-9-05966-055-7) pp 913–918

LOUDON, A G (1952) "The computation of permeability from simple soil tests" *Géotechnique*, vol 3, 4, Institution of Civil Engineers, London, pp 165–183

MAIR, R J, and WOOD, D M (1987) *Pressuremeter testing: methods and interpretation*, Butterworth-Heinemann Ltd, London (ISBN: 978-0-40802-434-1)

MANSUR, C I and KAUFMAN, R I (1962) "Dewatering". In: G A Leonards (ed) *Foundation engineering*, McGraw-Hill, New York, pp 241–350

MASTERS-WILLIAMS, H, HEAP, A, KITTS, H, GREENSHAW, L, DAVIS, S, FISHER, P, HENDRIE, M, OWENS, D (2001) *Control of water pollution from construction sites – guidance for consultants and contractors*, C532, CIRIA, London (ISBN: 978-0-86017-532-2). Go to: **www.ciria.org**

McNAMARA, A M, ROBERTS, T O L, MORRISON, P R J and HOLMES, G (2008) "Construction of a deep shaft for Crossrail" *Proceedings of the ICE – Geotechnical Engineering*, vol 161, 6, Institution of Civil Engineers, London, pp 299–309

McWHORTER, D B (1985) "Seepage in the unsaturated zone: a review". In: *A symposium of ASCE convention, Denver, Colorado, USA*, 5 May 1985, R L Volpe (ed), W E Kelly (eds) *Seepage and leakage from dams and impoundments*, Geotechnical Engineering Division, ASCE, New York, USA (ISBN: 978-0-87262-448-1), pp 200–219

MEIGH, A C (1987) *Cone Penetration Testing: methods and interpretation*, Butterworth-Heinemann, London (ISBN: 978-1-48310-860-5)

MILLER, E (1988) "The eductor dewatering system" *Proceedings of the ICE – Ground Engineering*, vol 21, 6, Institution of Civil Engineers, London, pp 29–34

MISSTEAR, B, BANKS, D and CLARK, L J (2006) *Water wells and boreholes*, Wiley, Chichester (ISBN 978-0-470-84989-7)

MURNANE, E, HEAP, A and SWAIN, A (2006) *Control of water pollution from linear construction projects. Technical guidance*, C648, CIRIA, London (ISBN: 978). Go to: **www.ciria.org**

NICHOLSON, D P (1994) "The observational method in geotechnical engineering: preface" *Géotechnique*, vol 44, 4, Institution of Civil Engineers, London, pp 613–618

NICHOLSON, D P, TSE, C-M and PENNY, C (1999) *The Observational Method in ground engineering: principles and applications*, R185, CIRIA, London (ISBN: 978-0-86017-497-4). Go to: **www.ciria.org**

NORBURY, D R (2010) *Soil and rock description in engineering practice*, Whittles Publishing, Caithness (ISBN: 978-1-90444-565-4)

NYER, E K (2009) *Groundwater treatment technology*, Wiley, UK (ISBN: 978-0-47165-742-2)

OVE ARUP & PARTNERS and GILBERTSON, A (2015a) *CDM2015 – construction work sector guidance for designers, fourth edition*, C755, CIRIA, London (ISBN: 978-0-86017-756-2). Go to: **www.ciria.org**

PECK, R B (1969a) "Advantages and limitations of the observational method in applied soil mechanics" *Géotechnique*, vol 19, 2, Institution of Civil Engineers, London, pp 171–187

PECK, R B (1969b) "Deep excavations and tunnelling in soft ground: state-of-art report". In: *Proc 7th int conf on soil mechanics and foundation engineering, Mexico City, Mexico*, pp 225–281

PERRY, J G, THOMPSON, P A and WRIGHT, M (1985) *Target and cost reimbursable construction contracts*, R85, CIRIA, London (ISBN: 978-0-86017-245-1). Go to: **www.ciria.org**

POTTER, M (1995) *Planning to build? A practical introduction to the construction process*, SP113, CIRIA, London (ISBN: 978-0-86017-433-2). Go to: **www.ciria.org**

POWERS, J P (1985) *Dewatering – avoiding its unwanted side effects*, American Society of Civil Engineers, New York, USA (ISBN: 978-0-87262-459-7)

POWERS, J P, CORWIN, A B, SCHMALL, P C and KAECK, W E (2007) *Construction dewatering and groundwater control: new methods and applications, third edition*, Wiley-Blackwell, New York, USA (ISBN: 978-0-47147-943-7)

POWRIE, W (2014) *Soil mechanics: concepts and applications, third edition*, CRC Press, Boca Raton, USA (ISBN: 978-1-46655-209-8)

POWRIE, W and PREENE, M (1992) "Equivalent well analysis of construction dewatering systems" *Géotechnique*, vol 42, 4, Institution of Civil Engineers, London, pp 635–639

POWRIE, W and PREENE, M (1994a) "Time–drawdown behaviour of construction dewatering systems in fine soils" *Géotechnique*, vol 44, 1, Institution of Civil Engineers, London, pp 83–100

POWRIE, W and PREENE, M (1994b) "Performance of ejectors in construction dewatering systems" *Proceedings of the ICE – Geotechnical Engineering*, vol 107, 3, Institution of Civil Engineers, London, pp 143–154

POWRIE, W and ROBERTS, T O L (1990) "Field trial of an ejector well dewatering system at Conwy, North Wales" *Quarterly Journal of Engineering Geology and Hydrogeology*, vol 23, 2, The Geological Society, London, pp 169–185

POWRIE, W and ROBERTS, T O L (1995) "Case history of a dewatering and recharge system in chalk" *Géotechnique*, vol 45, 3, Institution of Civil Engineers, London, pp 599–609

POWRIE, W, ROBERTS, T O L and JEFFERIS, S A (1990) "Biofouling of site dewatering systems". In: P Howsam, P (ed) *Microbiology in civil engineering: international proceedings of the Federation of European Microbiologial Societies symposium held at Cranfield Institute of Technology, Abingdon, UK*, Routledge, *UK*, 3–5 September 1990, E & F Spon, UK (ISBN: 0-41916-730-7), pp 341–352

POWRIE, W, ROBERTS, T O L and MOGHAZI H E-D (1989) "Effects of high permeability lenses on efficiency of wellpoint dewatering" *Géotechnique*, vol 39, 3, Institution of Civil Engineers, London, pp 543–547

PREENE, M (2000) "Assessment of settlements caused by groundwater control" *Proceedings of the ICE – Geotechnical Engineering*, vol 143, 4, Institution of Civil Engineers, London, pp 177–190

PREENE, M (2004) "Robert Stephenson (1803–59) – the first groundwater engineer". In: J D Mather (ed) *200 Years of British Hydrogeology*, Geological Society Special Publication 225, The Geological Society, London, pp 107–119

PREENE, M and BRASSINGTON, F C (2003) "Potential groundwater impacts from civil engineering works" *Water and Environmental Management Journal*, vol 17, 1, Wiley, London, pp 59–64

PREENE, M and POWRIE, W (1993) "Steady-state performance of construction dewatering systems in fine soils" *Géotechnique*, vol 43, 2, Institution of Civil Engineers, London, pp 191–205

PREENE, M and POWRIE, W (1994) "Construction dewatering in low permeability soils: some problems and solutions" *Proceedings of the ICE – Geotechnical Engineering*, vol 107, 1, Institution of Civil Engineers, London, pp 17–26

PREENE, M and ROBERTS, T O L (1994) "The application of pumping tests to the design of construction dewatering systems". In: *Proc of the int conf organised by the Institution of Civil Engineers, London*, 2–3 June 1993, W B Wilkinson (ed) *Groundwater problems in urban area*, Thomas Telford, London (ISBN: 978-0-72771-974-4) pp121–133

PRICE, M (1996) *Introducing groundwater*, Taylor and Francis, Abingdon, Oxon (ISBN: 978-0-74874-371-1)

PRIVETT, K D, MATTHEWS, S C and HODGES, R A (1996) *Barriers, liners and cover systems for containment and control of land contamination*, SP124, CIRIA, London (ISBN: 978-0-86017-437-0). Go to: **www.ciria.org**

PULLER, M and PULLER, D (2003) *Deep excavations: a practical manual, second edition*, Institution of Civil Engineers, London (ISBN: 978-0-72773-459-4)

QUINONES-ROZO, C (2010) "Lugeon test interpretation, revisited". In: *Collaborative Management of integrated watersheds, 30th annual USSD conference, Sacramento, California*, USA, 12–16 April 2010 (ISBN: 978-1-88457-551-8), pp 405–414

RABEN-LEVETZAU, J, MARKUSSEN, L M, BITSCH, K, NICOLAISEN, L L (2004) "Copenhagen Metro: groundwater control in sensitive urban areas". In: *Proc of the 30th ITA-AITES World Tunnel Congress*, Singapore, 22–27 May 2004. Special issue, *Tunnelling and underground space technology. Underground space for sustainable urban development*, Paper C04, Elseveier BV, UK

RAO, D B (1973) "Construction dewatering by vacuum wells" *Indian Geotechnical Journal*, vol 3, 3, Springer, India, pp 217–224

RAWLINGS, C G, HELLAWELL, E E and KILKENNY, W M (2000) *Grouting for ground engineering*, C514, CIRIA, London (ISBN: 978-0-86017-514-8). Go to: **www.ciria.org**

RIJKSWATERSTAAT (1986) *Groundwater infiltration with bored wells*, Rijkswaterstaat Communications 39, The Hague, The Netherlands

ROBERTS, T O L and DEED, M E R (1994) "Cost overruns in construction dewatering". In: B O Skipp (ed) *Risk and reliability in ground engineering*, Thomas Telford Publishing, London (ISBN: 978-0-72771-986-7)

ROBERTS, T O L, and HOLMES, G (2011) "Case study of a dewatering and recharge system in weak Chalk rock". In: *Proc 15th European conference on soil mechanics and geotechnical engineering*, Athens, Greece, 12–15 September 2011. A Anagnostopoulos, M Pachakis, C H Tsatsanifos (eds) *Geotechnics of hard soils – weak rocks (part 4)*, IOS Press, Amsterdam, The Netherlands (ISBN: 978-1-61499-199-1)

ROBERTS, T O L and PREENE, M (1994a) "Range of application of groundwater control systems". In: *Proc of the int conf organised by the Institution of Civil Engineers, London*, 2–3 June 1993, W B Wilkinson (ed) *Groundwater problems in urban area*, Thomas Telford, London (ISBN: 978-0-72771-974-4) pp 415–423

ROBERTS, T O L and PREENE, M (1994b) "The design of groundwater control systems using the observational method" *Géotechnique*, vol 44, 4, Institution of Civil Engineers, London, pp 727–734

ROBERTS, T O L, ROSCOE, H, POWRIE, W and BUTCHER, J E (2007) "Controlling clay pore pressures for cut-and-cover tunnelling" *Proceedings of the ICE – Geotechnical Engineering*, vol 160, 4, Institution of Civil Engineers, London, pp 227–236

ROBERTS, T O L, BOTHA, C P, and WELCH, A (2009) "Design and operation of a large dewatering system in Dubai". In: *Proc 17th int conf on soil mechanics and geotechnical engineering (ICSMGE), Alexandria, Egypt*, 5–9 October 2009, M Hamza, M Shahien, Y El-Mossallamy (eds) *The academia and practice of geotechnical engineering*, IOS Press, Amsterdam, The Netherlands, (ISBN: 978-1-60750-031-5)

ROBERTSON, P K and CAMPANELLA, R G (1983) "Interpretation of cone penetration tests. Part 1: Sand, Part 2: Clay" *Canadian Geotechnical Journal,* vol 20, 4, NRC Research Press, Canada, pp 718–745

ROWE, P W (1968) "Failure of foundation and slopes in layered deposits in relation to site investigation" *Proceedings of the Institution of Civil Engineers*, vol 39, 4, Institution of Civil Engineers, London, pp 73–131

ROWE, P W (1972) "The relevance of soil fabric to site investigation practice" *Géotechnique*, vol 22, 2, Institution of Civil Engineers, London, pp 195–300

ROWE, P W (1986) "The potentially latent dominance of groundwater in ground engineering". In: J C Cripps, F G Bell and M G Culshaw (eds) *Groundwater in Engineering Geology*, Geological Society Special Publication 3, The Geological Society, London, pp 27–42

RUSHTON, K R and REDSHAW, S C (1979) *Seepage and groundwater flow: numerical analysis by analog and digital methods*, Wiley, Chicester (ISBN: 978-0-47199-754-2)

SHERARD, J L, DUNNIGAN, L P and TALBOT, J R (1984a) "Basic properties of sand and gravel filters" *ASCE Journal of Geotechnical Engineering*, vol 110, 6, American Society of Civil Engineers, Reston, USA, pp 684–700

SHERARD, J L, DUNNIGAN, L P and TALBOT, J R (1984b) "Filters for silts and clays" *ASCE Journal of Geotechnical Engineering*, vol 110, 6, American Society of Civil Engineers, Reston, USA, pp 701–718

SITE INVESTIGATION STEERING GROUP (1993) *Site investigation in construction*
*Part 1 Without investigation ground is a hazard* (ISBN: 978-0-72771-982-9)
*Part 2 Planning, procurement and quality management* (ISBN: 978-0-72771-983-6)
*Part 3 Specification for ground investigation* (978-0-72771-984-3)
*Part 4 Guidelines for the safe investigation by drilling of landfills and contaminated land* (ISBN: 978-0-72771-985-0)
Institution of Civil Engineers, London, UK

SITE INVESTIGATION STEERING GROUP (2011) *UK Specification for ground investigation, second edition*, Thomas Telford Ltd, London (ISBN: 978-0-72773-506-5)

SITE INVESTIGATION STEERING GROUP (2013) *Effective site investigation (site investigation in construction series)*, Thomas Telford, London (ISBN: 978-0-72773-505-8)

SIWEC, T M and WHITE, J K (1995) "A design procedure for multi-jet pump installation" *Proceedings of the ICE – Water Maritime and Energy*, vol 112, 4, Institution of Civil Engineers, London pp 304–315

SLOCOMBE, R, BUCHANAN, J and LAMONT, D (2003) *Engineering and health in compressed air work*, Thomas Telford Publishing, London (ISBN: 978-0-72774-011-3)

STERRETT, R (2009) *Groundwater and wells, third edition*, Johnson Division, St. Paul, Minnesota, USA (ISBN: 978-0-97877-930-6)

STROUD, M A (1987) "Groundwater control – general report". In: *Proc of the 9th conf on soil mechanics and foundation engineering, Dublin, Ireland*, 31 August 1987. E T Hanrahan, T L L Orr, T F Widdis (eds) *Groundwater effects in geotechnical engineering*, vol 1–3, Balkema, Rotterdam, pp 983–1008

TERZAGHI, K, PECK, R B and MESRI, G (1996) *Soil mechanics in engineering practice*, John Wiley & Sons, New York, USA (ISBN: 978-0-47108-658-1)

THEIS, C V (1935) "The relation between the lowering of the piezometric surface and the rate and duration of discharge of a well using groundwater storage" *Transactions of the American Geophysical Union*, vol 16, 2, American Geophysical Union, USA, pp 519–524

TROUGHTON, V M (1987) "Groundwater control by pressure relief and recharge". In: *Proc of the 9th conf on soil mechanics and foundations, Dublin, Ireland*, 31 August 3 September 1987, E T Hanrahan, T L L Orr, T F Widdis (eds) *Groundwater effects in geotechnical engineering*, CRC Press, Boca Raton, USA (ISBN: 978-9-06191-720-4), pp 259–264

WALTHALL, S and CAMPBELL, J E (1986) "The measurement and use of permeability values with specific reference to fissured aquifers". In: J C Cripps, F G Bell and M G Culshaw (eds) *Groundwater in Engineering Geology*, Geological Society Special Publication 3, The Geological Society, London, pp 273–278

WELTMAN, A J and HEAD, J M (1983) *Site investigation manual*, SP25, CIRIA, London (ISBN: 978-0-86017-196-6). Go to: **www.ciria.org**

WESTCOTT, F J, LEAN, C M B and CUNNINGHAM, M L (2001) *Piling and penetrative ground improvement methods on land affected by contamination: guidance on pollution protection*, Report NC/99/73, National Groundwater and Contaminated Land Centre, Environment Agency, Solihull, UK

WILD, J L and MONEY, M S (1986) "Results of an experimental programme of in-situ permeability testing in rock". In: J C Cripps, F G Bell and M G Culshaw (eds) *Groundwater in engineering geology*, Geology Special Publication 3, The Geological Society, London, pp 283–293

WILLIAMS, B P and WAITE, D (1993) *The design and construction of sheet-piled cofferdams*, SP95, CIRIA, London (ISBN: 978-0-86017-361-8). Go to: **www.ciria.org**

YOUNGER, P L (2007) *Groundwater in the environment: an introduction*, Blackwell Publishing, Oxford, UK (ISBN: 978-1-40512-143-9)

YUNGWIRTH, G, PREENE, M, DOBR, M and FORERO GARCIA, F (2013) "Practical application and design considerations for fully grouted vibrating wire piezometers in minewater investigations". In: *Annual International Mine Water Association Conference 2013, Golden, Colorado, USA*, 6–9 August 2013, A Brown, L Figueroa, C Wolkersdorfer (eds) *Reliable mine water technology, vol 1*, Curran Associates Inc, New York, USA (ISBN 978-0-61579-385-6) pp 229–237

## Statutes

### Acts

Health and Safety at Work Act 1974 (c.37)

Water Resources Act 1991 (c.57)

### Directives

Directive 2000/60/EC of the European Parliament and of the Council of 23 October 2000 establishing a framework for Community action in the field of water policy (Water Framework Directive)

### Orders

Water (Northern Ireland) Order 1999 (No.662/NI 6)

### Regulations

Construction (Head Protection) Regulations 1989 (No.2209)

Construction (Design and Management) Regulations 2015 (CDM2015) (No.51)

The Control of Noise at Work Regulations 2005 (No.1643)

Electricity at Work Regulations 1989 (No.635)

The Environmental Permitting (England and Wales) Regulations 2010

Fire Precautions Act 1971 (c.40)

The Lifting Operations and Lifting Equipment Regulations (LOLER) 1998 (No.2307)

The Management of Health and Safety at Work Regulations 1992 (No.3242)

The Personal Protective Equipment at Work Regulations 1992 (No.2966)

The Provision and Use of Work Equipment Regulations (PUWER) 1998 (No.2306)

Work at Height Regulations 2005 (No.735)

Water Abstraction and Impoundment (Licensing) Regulations (Northern Ireland) 2006 (No.482)

The Water Environment (Controlled Activities) (Scotland) Regulations 2011 (No.209)

## Standards

### British

BS 5930:2015 *Code of practice for ground investigations*

BS 7022:1988 *Geophysical logging of boreholes for hydrogeological purposes*

BS 7671:2008 *Requirements for electrical installations: IEE Wiring Regulations 17th edition*

BS EN 1536:2010 *Execution of special geotechnical works. Bored piles*

BS EN 1538:2010 *Execution of special geotechnical works. Diaphragm walls*

BS EN 1997-1:2004 *Eurocode 7: Geotechnical design. Part 1: General rules*

BS EN 1997-2:2007 *Eurocode 7: Geotechnical design. Part 2: Ground investigation and testing*

BS EN 12063:1999 *Execution of special geotechnical works – sheet pile walls*

BS EN 12715:2000 *Execution of special geotechnical works. Grouting*

BS EN 12716:2001 *Execution of special geotechnical works. Jet grouting*

BS EN 14688-1:2002 *Geotechnical investigation and testing – identification and classification of soil. Part 1: Identification and description*

BS EN 14689-1:2003 *Geotechnical investigation and testing – identification and classification of rock. Part 1: Identification and description*

BS EN ISO 17892-2:2014 *Geotechnical investigation and testing. Laboratory testing of soil. Determination of bulk density*

BS EN ISO 22282-1:2012 *Geotechnical investigation and testing. Geohydraulic testing. General rules*

BS EN ISO 22282-2:2012 *Geotechnical investigation and testing. Geohydraulic testing. Water permeability tests in a borehole using open systems*

BS EN ISO 22282-3:2012 *Geotechnical investigation and testing. Geohydraulic testing. Water pressure tests in rock*

BS EN ISO 22282-4:2012 *Geotechnical investigation and testing. Geohydraulic testing. Pumping tests*

BS EN ISO 22282-5:2012 *Geotechnical investigation and testing. Geohydraulic testing. Infiltrometer tests*

BS EN ISO 22282-6:2012 *Geotechnical investigation and testing. Geohydraulic testing. Water permeability tests in a borehole using closed systems*

BS ISO 1438:2008 *Hydrometry: open channel flow measurement using thin-plate weirs*

BS ISO 5667-11:2009 *Water quality: sampling. guidance on sampling of groundwaters*

BS ISO 14686:2003 *Hydrometric determinations. Pumping tests for water wells. Considerations and guidelines for design, performance and use*

# Further reading


BLACK, J H (2010) "The practical reasons why slug tests (including falling and rising head tests) often yield the wrong value of hydraulic conductivity" *Quarterly Journal of Engineering Geology and Hydrogeology*, vol 43, August, The Geological Society, London, pp 345–358

DAVIS, G M and HORSWILL, P (2002) "Groundwater control and stability in an excavation in Magnesian Limestone near Sunderland, NE England" *Engineering Geology*, vol 66, 1–2, Elsevier BV, UK, pp 1–18

DEED, M E R and PREENE, M (2015) "Managing the clogging of water wells". In: *Proc of the XVI ECSMFGE Geotechnical engineering for infrastructure and development, Edinburgh, Scotland,* 13–17 September 2015. M G Winter, D M Smith, P J L Eldred and D G Toll (2015) *Problematic materials, environment water and energy*, ICE Publishing, London, pp 2787–2792

HARTWELL, D J (2015) "Permeability testing problems in rock". In: *Proc of the XVI ECSMFGE Geotechnical engineering for infrastructure and development, , Edinburgh, Scotland,* 13–17 September 2015. M G Winter, D M Smith, P J L Eldred and D G Toll (2015) *Problematic materials, environment water and energy*, ICE Publishing, London, pp3657–3662

LAWRENCE, U, BEAGLEY, R, NORGATE, S and THROWER, A (2015) "Variation in permeability and dewatering performance for part of the Crossrail route in east London". In: *Proc of the XVI ECSMFGE Geotechnical engineering for infrastructure and development, , Edinburgh, Scotland,* 13–17 September 2015. M G Winter, D M Smith, P J L Eldred and D G Toll (2015) *Problematic materials, environment water and energy*, ICE Publishing, London, pp849–854

LINNEY, L F and WITHERS, A D (1998) "Dewatering the Thanet beds in SE London: three case histories" *Quarterly Journal of Engineering Geology and Hydrogeology*, vol 31, 2, The Geological Society, London, pp 115–122

LONG, M, MURPHY, M, ROBERTS, T O L, O'BRIEN, J and CLANCY, N (2015) "Deep excavations in water-bearing gravels in Cork", *Quarterly Journal of Engineering Geology and Hydrogeology*, 48, 2, The Geological Society, London, pp79–93

PREENE, M and FISHER, S (2015) "Impacts from groundwater control in urban areas". In: *Proc of the XVI ECSMFGE Geotechnical engineering for infrastructure and development, Edinburgh, Scotland,* 13–17 September 2015. M G Winter, D M Smith, P J L Eldred and D G Toll (2015) *Problematic materials, environment water and energy*, ICE Publishing, London, pp2846–2852

PREENE, M and LOOTS, E (2015) "Optimisation of dewatering systems". In: *Proc of the XVI ECSMFGE Geotechnical engineering for infrastructure and development, Edinburgh, Scotland,* 13–17 September 2015. M G Winter, D M Smith, P J L Eldred and D G Toll (2015) *Problematic materials, environment water and energy*, ICE Publishing, London, pp2841–2846

PREENE, M and ROBERTS, T O L (2002) "Groundwater control for construction in the Lambeth Group" *Proceedings of the ICE – Geotechnical Engineering*, vol 155, 4, Institution of Civil Engineers, London, pp 221–227

ROBERTS, T O L, LINDE, E and SUTTON, M (2015) "In-tunnel dewatering for a cross passage in London". In: *Proc the third Arabian Tunnelling Conference, Dubai, United Arab Emirates*, 23–25 November 2015

ROBERTS, T O L, LINDE, E, VINCENTE, C and HOLMES, G (2015) "Multi-aquifer pressure relief in East London". In: *Proc of the XVI ECSMFGE Geotechnical engineering for infrastructure and development, Edinburgh, Scotland,* 13–17 September 2015. M G Winter, D M Smith, P J L Eldred and D G Toll (2015) *Problematic materials, environment water and energy*, ICE Publishing, London, pp2811–2816

ROBERTS, T O L, SMITH, R, STÄRK, A and ZEISZIG, W (2015) "Sub-surface dewatering for an inclined SCL tunnel". In: *Proc of the XVI ECSMFGE 2015 Geotechnical engineering for infrastructure and development, Edinburgh, Scotland,* 13–17 September 2015. M G Winter, D M Smith, P J L Eldred and D G Toll (2015) *Problematic materials, environment water and energy*, ICE Publishing, London, pp 2853–2858

WHITAKER, D (2004) "Groundwater control for the Stratford CTRL station box", *Proceedings of the ICE – Geotechnical Engineering*, vol 157, 4, Institution of Civil Engineers, London, pp 183–191

WOODWARD, J (2005) *An introduction to geotechnical processes*, Spon Press, Abingdon, Oxford, (ISBN: 978-0-41528-646-6)

# A1 Datasheets

## DATASHEET 1: CONVERSION FACTORS FOR UNITS

Example: to convert 10 miles to kilometres, find 1 mile in the 'length' table. Values on a horizontal row are equal, eg 1 mile = 1.609 km, therefore 10 miles = 16.09 km.

**Length**

| *km* | *m* | *mm* | *mile* | *yard* | *ft* | *in* |
|---|---|---|---|---|---|---|
| 1 | 1000 | 106 | 0.6214 | 1094 | 3281 | $3.94 \times 10^{4}$ |
| $10^{-3}$ | 1 | 1000 | $6.21 \times 10^{-4}$ | 1.094 | 3.281 | 39.370 |
| $10^{-6}$ | $10^{-3}$ | 1 | $6.21 \times 10^{-7}$ | $1.09 \times 10^{-3}$ | $3.28 \times 10^{-3}$ | 0.0394 |
| 1.609 | 1609.4 | $1.61 \times 10^{6}$ | 1 | 1760 | 5280 | 63360 |
| $9.14 \times 10^{-4}$ | 0.9144 | 914.41 | $5.68 \times 10^{-4}$ | 1 | 3 | 36 |
| $3.05 \times 10^{-4}$ | 0.3048 | 304.8 | $1.89 \times 10^{-4}$ | 0.3333 | 1 | 12 |
| $2.54 \times 10^{-5}$ | 0.0254 | 25.4 | $1.58 \times 10^{-5}$ | $2.78 \times 10^{-2}$ | $8.33 \times 10^{-2}$ | 1 |

**Volume**

| *$m^3$* | *litre* | *$cm^3$ (ml)* | *$yd^3$* | *$ft^3$* | *$in^3$* | *gallon* | *US gallon* |
|---|---|---|---|---|---|---|---|
| 1 | $10^{3}$ | $10^{6}$ | 1.308 | 35.311 | 61013 | 219.97 | 264.17 |
| $10^{-3}$ | 1 | 103 | $1.31 \times 10^{-3}$ | $3.53 \times 10^{-2}$ | 61.013 | 0.220 | 0.2642 |
| $10^{-6}$ | $10^{-3}$ | 1 | $1.31 \times 10^{-6}$ | $3.53 \times 10^{-5}$ | $6.10 \times 10^{-2}$ | $2.20 \times 10^{-4}$ | $2.64 \times 10^{-4}$ |
| 0.7646 | 764.6 | $7.65 \times 10^{5}$ | 1 | 27 | 46650 | 168.19 | 201.98 |
| $2.83 \times 10^{-2}$ | 28.32 | $2.831 \times 10^{4}$ | $3.70 \times 10^{-2}$ | 1 | 1728 | 6.229 | 7.481 |
| $1.64 \times 10^{-5}$ | $1.64 \times 10^{-2}$ | 16.39 | $2.14 \times 10^{-5}$ | $5.79 \times 10^{-4}$ | 1 | $3.61 \times 10^{-3}$ | 0.00433 |
| $4.55 \times 10^{-3}$ | 4.546 | $4.55 \times 10^{3}$ | $5.95 \times 10^{-3}$ | 0.161 | 277.42 | 1 | 1.201 |
| $3.79 \times 10^{-3}$ | 3.785 | 3785.4 | $4.95 \times 10^{-3}$ | 0.134 | 230.96 | 0.8327 | 1 |

**Flow rate**

| *l/s* | *l/min* | *$m^3$/s* | *$m^3$/hr* | *$m^3$/day* | *gallon/min* | *gallon/hr* | *gallon/day* |
|---|---|---|---|---|---|---|---|
| 1 | 60 | 0.001 | 3.6 | 86.4 | 13.2 | 792 | $1.9 \times 10^{4}$ |
| 0.0167 | 1 | $1.67 \times 10^{-5}$ | 0.06 | 1.44 | 0.22 | 13.2 | 316.8 |
| 1000 | $6.00 \times 10^{4}$ | 1 | 3600 | 86400 | $1.32 \times 10^{4}$ | $7.92 \times 10^{5}$ | $1.9 \times 10^{7}$ |
| 0.0115 | 0.694 | $1.16 \times 10^{-5}$ | 0.0417 | 1 | 0.153 | 9.167 | 220 |
| 0.0758 | 4.55 | $7.58 \times 10^{-5}$ | 0.273 | 6.546 | 1 | 60 | 1440 |
| 0.00126 | 0.0758 | $1.26 \times 10^{-6}$ | 0.00455 | 0.1091 | 0.0167 | 1 | 24 |
| $5.26 \times 10^{-5}$ | $3.16 \times 10^{-3}$ | $5.26 \times 10^{-8}$ | $1.89 \times 10^{-4}$ | 0.00455 | $6.94 \times 10^{-4}$ | 0.0417 | 1 |

| Pressure and head | | | | | | | |
|---|---|---|---|---|---|---|---|
| *m $H_2O$* | *ft $H_2O$* | *in Hg* | *mm Hg* | *Bar* | *psi (lb/in²)* | *kPa (kN/m²)* | *Atmosphere* |
| 1 | 3.281 | 2.896 | 73.55 | 0.0979 | 1.42 | 9.789 | 0.0966 |
| 0.3048 | 1 | 0.883 | 22.42 | 0.0299 | 0.433 | 2.984 | 0.0295 |
| 0.345 | 1.133 | 1 | 25.4 | 0.0339 | 0.491 | 3.381 | 0.0334 |
| 0.0136 | 0.0446 | 0.0394 | 1 | 0.00133 | 0.0193 | 0.133 | 0.00132 |
| 10.204 | 33.480 | 29.551 | 750 | 1 | 14.51 | 100 | 0.989 |
| 0.703 | 2.307 | 2.036 | 51.72 | 0.0689 | 1 | 6.884 | 0.0682 |
| 0.102 | 0.335 | 0.296 | 7.513 | 0.01 | 0.1451 | 1 | 0.0099 |
| 10.309 | 33.825 | 29.856 | 760 | 1.01 | 14.66 | 100.93 | 1 |

| Permeability | | | | | |
|---|---|---|---|---|---|
| *m/s* | *cm/s* | *m/year* | *m/day* | *ft/yr* | *ft/day* |
| 1 | 100 | $3.16 \times 10^{7}$ | $8.64 \times 10^{4}$ | $1.04 \times 10^{8}$ | $2.83 \times 10^{5}$ |
| 0.01 | 1 | $3.16 \times 10^{5}$ | 864 | $1.04 \times 10^{6}$ | $2.83 \times 10^{3}$ |
| $3.17 \times 10^{-8}$ | $3.17 \times 10^{-6}$ | 1 | 0.00274 | 3.281 | $8.98 \times 10^{-3}$ |
| $1.16 \times 10^{-5}$ | 0.00116 | 365.25 | 1 | 1198.36 | 3.281 |
| $9.66 \times 10^{-9}$ | $9.66 \times 10^{-7}$ | 0.305 | $8.34 \times 10^{-4}$ | 1 | $2.74 \times 10^{-3}$ |
| $3.53 \times 10^{-6}$ | $3.53 \times 10^{-4}$ | 111.32 | 0.305 | 365.25 | 1 |

# DATASHEET 2: FRICTION LOSSES IN PIPEWORK

## Friction losses in header and discharge pipes

Note: friction head loss may be estimated by assuming that the total output from the wellpoints flows the full length of the header pipe.

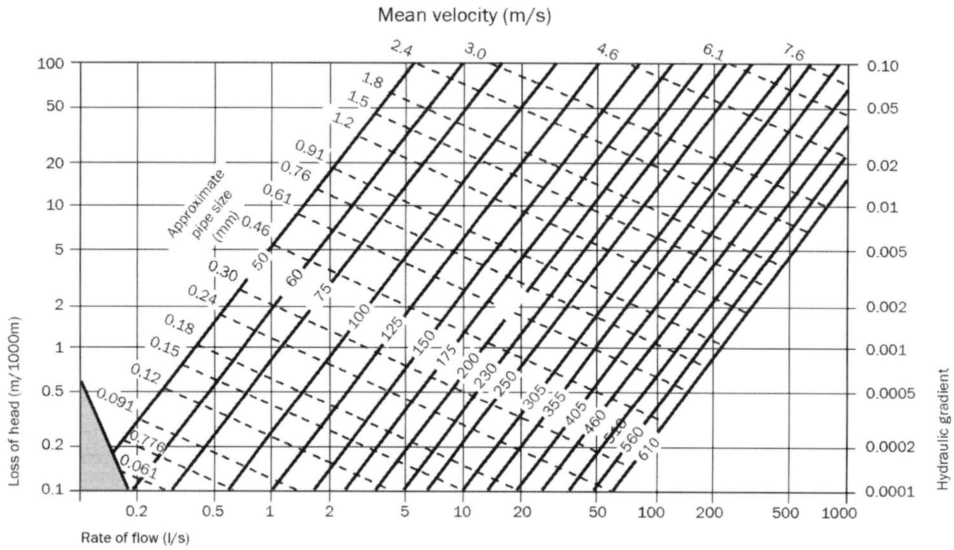


*Table A1.1 Friction losses in valves and fittings as an equivalent length of straight pipe in metres*

| Type of fitting 150 | | Nominal pipe diameter (mm) | | | | | | | | |
|---|---|---|---|---|---|---|---|---|---|---|
| | | 150 | 200 | 250 | 300 | 450 | 600 | 750 | 1065 | 1200 |
| Gate valve | Open | 1.1 | 1.4 | 1.7 | 2.0 | 2.8 | 4.3 | 5.2 | 6.4 | 7.6 |
| | ¼ closed | 6.1 | 7.9 | 10.1 | 12.2 | 18.3 | 24.4 | 30.5 | 41.2 | 48.8 |
| | ½ closed | 30.5 | 39.6 | 51.8 | 59.5 | 91.5 | 122.0 | 152.0 | 213.0 | 244.0 |
| | ¾ closed | 122.0 | 159.0 | 213.0 | 244.0 | 366.0 | 488.0 | 610.0 | 854.0 | 976.0 |
| Standard tree | Flow in-line | 2.9 | 4.3 | 5.0 | 5.9 | 9.1 | 11.9 | 15.1 | 22.0 | 24.7 |
| | Flow to/ from branch | 9.8 | 12.8 | 16.8 | 19.8 | 30.5 | 39.6 | 50.3 | 73.2 | 82.3 |
| Medium sweep 90 elbow | | 4.3 | 5.5 | 6.7 | 7.9 | 12.2 | 15.9 | 21.3 | 28.0 | 32.0 |
| Long sweep 90 elbow | | 3.2 | 4.3 | 5.3 | 6.1 | 9.1 | 12.2 | 15.2 | 21.3 | 24.4 |
| Square 90 elbow | | 9.8 | 12.8 | 16.8 | 19.8 | 30.5 | 39.6 | 50.3 | 73.2 | 82.3 |
| 45 elbow | | 2.3 | 3.1 | 3.7 | 4.6 | 6.4 | 8.5 | 10.7 | 15.2 | 18.3 |
| Sudden enlargement (→d →D→) | d/D = ¼ | 4.9 | 6.4 | 8.4 | 9.9 | 15.2 | 19.8 | 25.2 | 36.6 | 41.2 |
| | d/D = ½ | 3.2 | 4.3 | 5.3 | 6.1 | 9.1 | 12.2 | 15.2 | 21.3 | 24.4 |
| | d/D = ¾ | 2.9 | 3.7 | 4.9 | 5.6 | 8.4 | 11.0 | 13.7 | 19.8 | 22.9 |
| Sudden contraction (→D →d→) | d/D = ¼ | 2.3 | 3.1 | 3.7 | 4.6 | 6.4 | 8.5 | 10.7 | 15.2 | 18.3 |
| | d/D = ½ | 1.7 | 2.3 | 2.9 | 3.4 | 4.9 | 6.4 | 8.2 | 11.3 | 12.8 |
| | d/D = ¾ | 1.1 | 1.4 | 1.7 | 2.0 | 2.8 | 4.3 | 5.2 | 6.4 | 7.6 |

# DATASHEET 3: V-NOTCH WEIR DISCHARGE CHARTS

Charts based on the methods of BS ISO 1438:2008. The depth of water, *h*, over the weir is measured above base of V-notch (see Box 3.3). The position of measurement should be upstream from the weir plate by a distance of approximately 1.1 m to 0.7 m, but not near a baffle or in the corner of a tank.

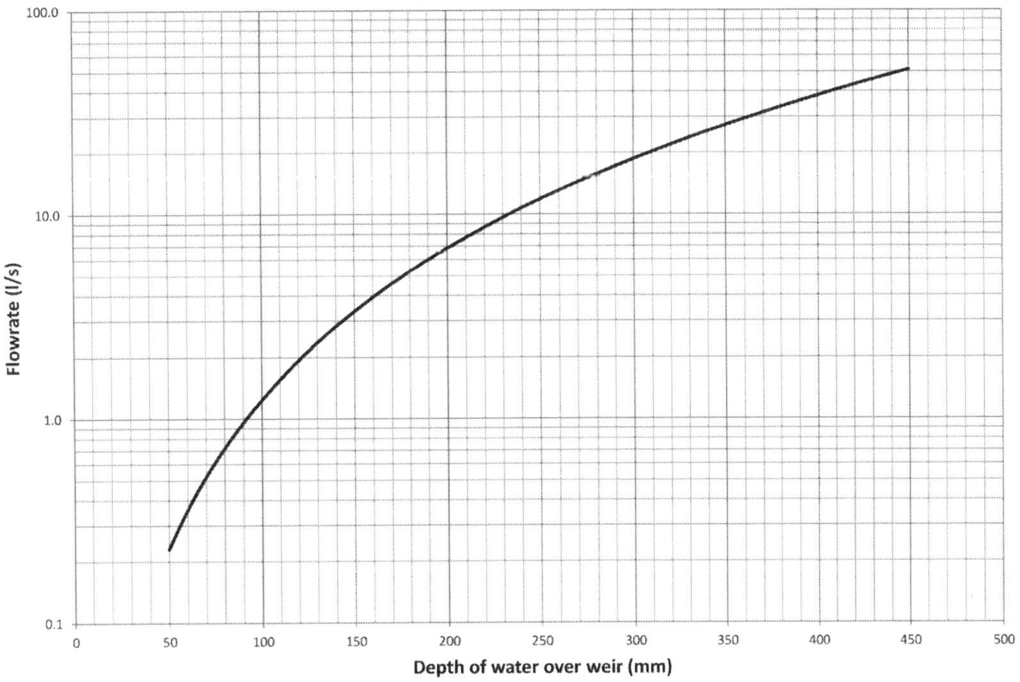


*Discharge chart for 30° V-notch weir*

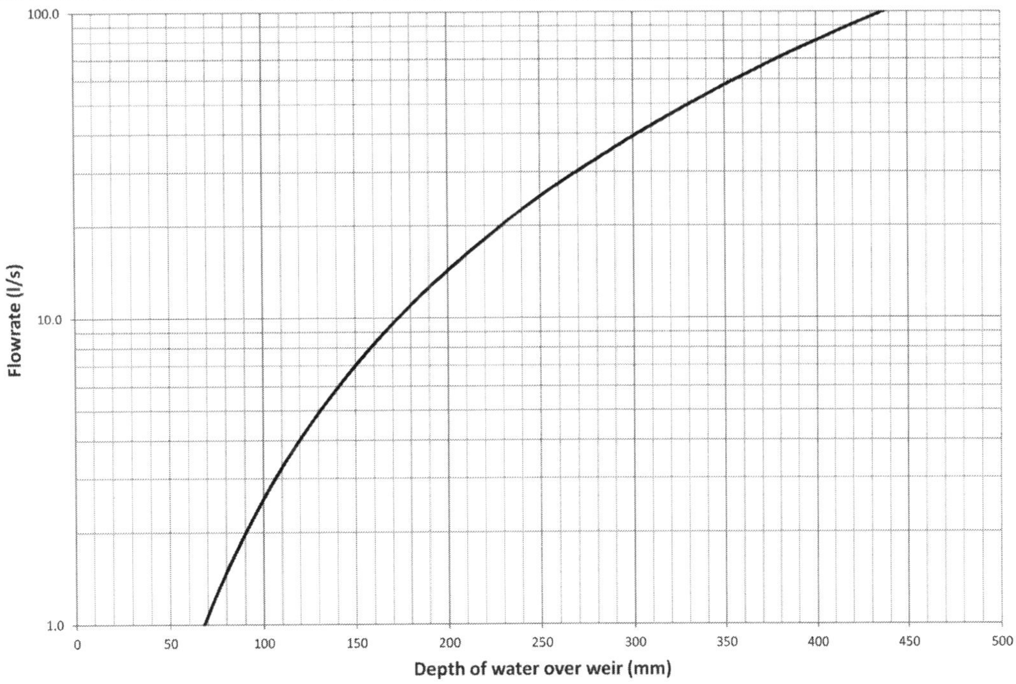


*Discharge chart for 60° V-notch weir*

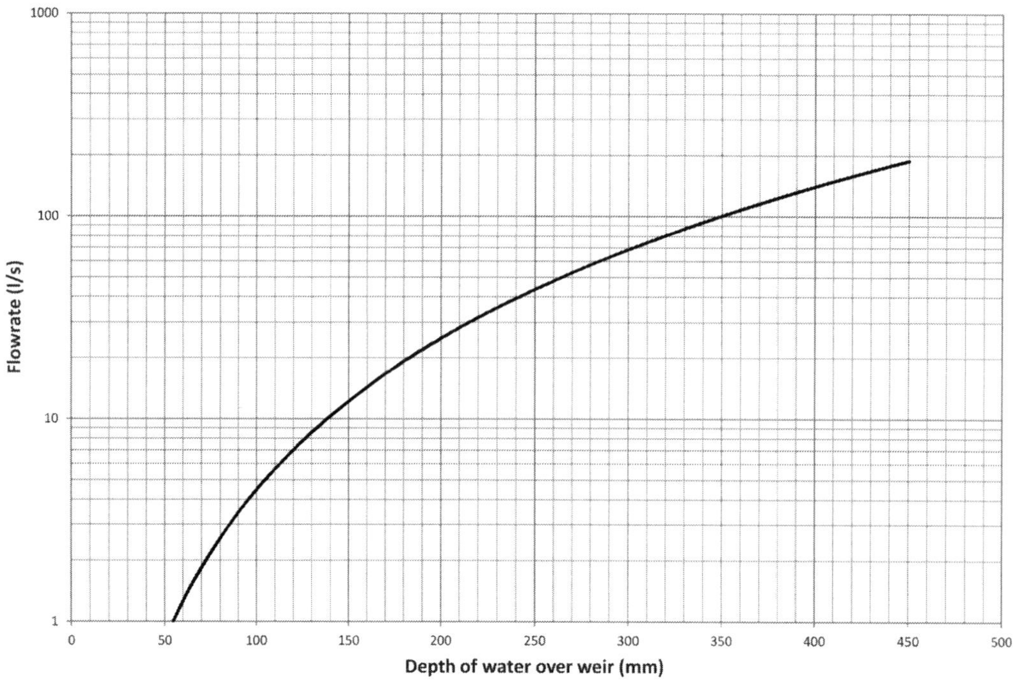


*Discharge chart for 90° V-notch weir*

# DATASHEET 4: PRUGH METHOD OF ESTIMATING PERMEABILITY OF SOILS (AFTER POWERS *ET AL*, 2007)

Permeability is estimated from the $D_{50}$ particle size, uniformity coefficient $U$ (where $U = D_{60}/D_{10}$) and the relative density of the soil using the diagrams below, interpolating as necessary.

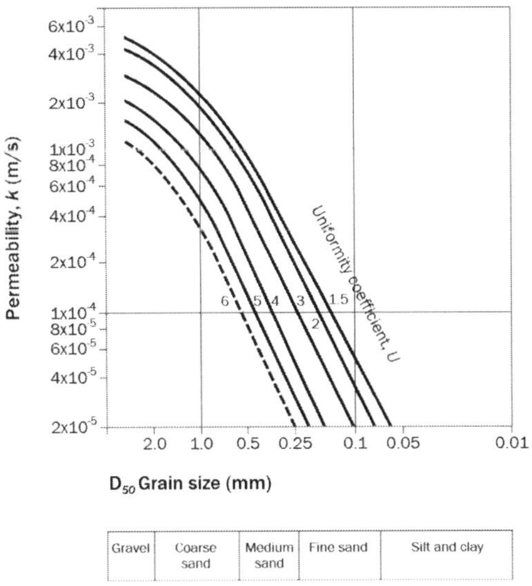


*a Dense soils*

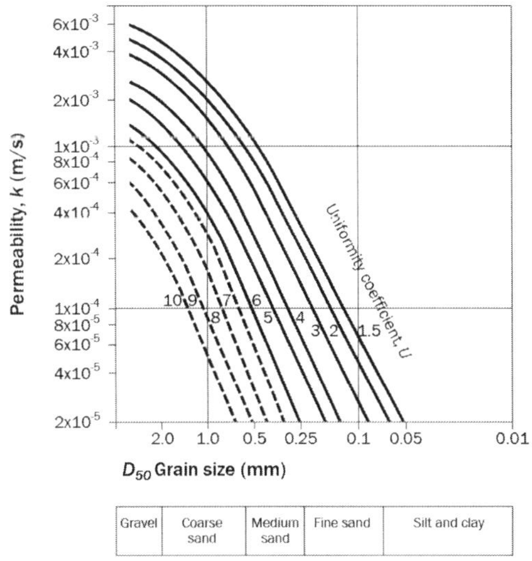


*b Medium dense soils*

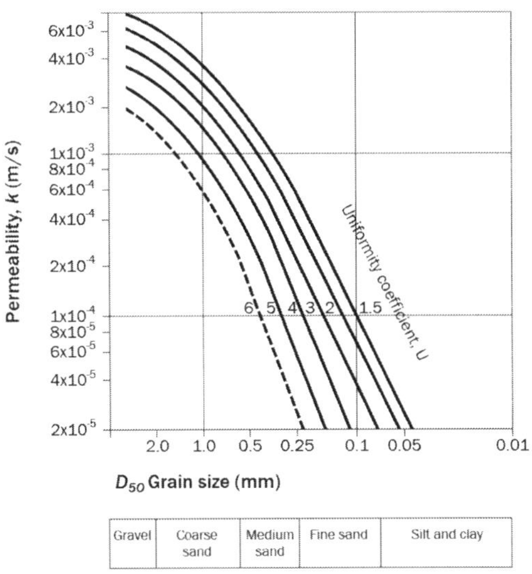


*c Loose soils*